INDUSTRIALIZATION OF SPACE IN THE 1990s

PROCEEDINGS OF
A ROYAL SOCIETY DISCUSSION MEETING
HELD ON 7 AND 8 DECEMBER 1983

ORGANIZED BY
SIR HARRIE MASSEY, F.R.S., AND G. K. C. PARDOE
AND EDITED BY
G. K. C. PARDOE

LONDON
THE ROYAL SOCIETY
1984

Printed in Great Britain for the Royal Society
at the
University Press, Cambridge

ISBN 0 85403 233 9

First published in *Philosophical Transactions of the Royal Society of London*,
series A, volume 312 (no. 1519), pages 1–140

Published by the Royal Society
6 Carlton House Terrace, London SW1Y 5AG

PREFACE

There are many fascinating and important applications of science and technology within the medium of space. The early motivations of scientific research have led, within a generation, to a deep and widespread involvement of projects and services operated within the space medium, now of benefit to so many people on Earth. There are several examples where the transition from basic science has fully matured into commercial application and this process is accelerating and broadening all the time. To those people professionally engaged in activities in space, the provision of equipment on one hand and the use and sale of services on the other, is a logical and visible evolution of increasing momentum. The full extent of that process, however, is by no means apparent to the world at large and the purpose of this meeting was to review the state that had been reached in this industrialization of space, and even more particularly to discuss the prospects and potential for the future.

This publication brings together the papers presented within a two day meeting held on 7 and 8 December 1983. Not for the first time, the Royal Society provided a unique forum for speakers of senior calibre from all over the world to report on their work and their ideas for the future. We were grateful to the Minister of State for Industry and Information technology, Mr Kenneth Baker, for opening the meeting on such a positive note and for emphasizing the importance that a proper involvement in this commercial process in space represents to the British community. Subsequent speakers from different walks of life underlined the ways in which such benefits can be taken up.

The organizers had some difficulty in formulating the programme, to judge whether the European Spacelab would have flown by the time of the meeting, as clearly this would have had a bearing on the content of some of the papers chosen. What could not have been foreseen was that Spacelab would in fact be in orbit precisely at the time of the symposium. We were therefore able to enjoy live reports on its progress and (as has certainly turned out) its achievements. Such a project underlines European expansion into the manned space field, and reflects the level of technological competence reached in this sector, as well as a variety of unmanned space activity. The full spectrum of activity in space and ground segments was covered fully in the meeting; the field is so great that priorities of interest and opportunity for individual companies and organizations – and indeed countries – must be carefully evaluated, and it is the hope that the opportunities outlined in the papers of this meeting will assist such a process.

There is one final aspect, which cannot be over-emphasized. It is the fact that space applications have already permeated laterally into many of our normal and orthodox industries and businesses; some of which could not now operate without the services so provided. It is, however, not realized by many, just how widespread *is* the influence of space activity. Britain has already demonstrated good capability in this field, but many of its industries and businesses, and some government departments, have not yet moved into the active role that the space situation permits and, indeed, demands. It is hoped that the papers provided in this publication will be helpful to professionals and entrepreneurs alike in advancing that situation.

A special word of thanks must be directed to all the Royal Society staff involved, and particularly to Miss Christine Johnson, whose efficient and friendly efforts in administering the conference were much appreciated by the organizers and were vital to its success.

April 1984   G. K. C. PARDOE

CONTENTS

[Two plates]

Phil. Trans. R. Soc. Lond. A **312**, 3 (1984)

Printed in Great Britain

[3]

Chairman's introduction

By R. Wilson, F.R.S.

*Department of Physics and Astronomy, University College London, Gower Street,
London WC1E 6BT, U.K.*

Although I consider it a privilege to open this meeting and to act as your Chairman for this first day, I do so with considerable sadness. I deputize for Sir Harrie Massey, F.R.S., who died at his home only eleven days ago after several months of illness. I happen to know how committed he was to this meeting, the great importance he attached to its subject matter and how very, very much he wanted to be here today. This is not the moment to outline his many achievements but I am sure you will allow me a brief, personal tribute, to which you may wish to subscribe. No one contributed more to the early development of space activities in this country or, indeed, in Europe than Harrie Massey. More than anyone else he provided the leadership and the inspiration. We will all miss him.

I must now turn to my introduction to this meeting, and I would like to say how appropriate I think it is that the Royal Society should be the host of a meeting on the subject of The Industrialization of Space. British space research was initiated in the aftermath of Sputnik and the American commitment to a space programme. It was then scientifically oriented, and was led, funded and managed by the Royal Society via its National Committee for Space Research, which was chaired by Harrie Massey, to whom I have just referred. British Industry cut its 'space teeth' on this programme, which included, for example, the U.K. series of Aerial satellites, and this gave it a competitive edge in seeking international as well as national space contracts. In addition, U.K. hardware, funded for science but involving British Industry, was flown in many N.A.S.A. and E.S.A. satellites such as N.A.S.A's Copernicus and Nimbus satellites, E.S.A's TD-1 and the N.A.S.A.–E.S.A.–S.E.R.C. satellite IUE; there are many other examples. These contributions, although minor in terms of resources, were major intellectually because of the degree of know-how and skill that was deployed. The returns to the U.K. were also major.

Since those early days, the whole pattern of space research has changed and expanded dramatically into many areas of applications and commerce. The extent to which this has already happened, together with the exciting possibilities for the future, is the subject of this meeting on the industrialization of space, for which the organizers have assembled a very distinguished list of speakers. It is very appropriate that the first speaker should be someone who is personally committed to the aim expressed in that title and who is the Minister responsible for it in the U.K., namely the Rt Hon. Kenneth Baker, Minister of State for Industry and Information Technology.

Phil. Trans. R. Soc. Lond. A **312**, 5–7 (1984)

Printed in Great Britain

[5]

Opening address

By K. Baker

Minister of State for Industry and Information Technology

Department of Trade and Industry, Ashdown House, 123 Victoria Street, London SW1, U.K.

I very much welcome this opportunity to provide this opening address.

It is only a little over 25 years ago that we were able to look out and see a bright new star, Sputnik 1, moving across the sky. At the time, it was just a miracle of technology and few people realized that mankind had been given the key to a whole new resource like that of the land, sea and air.

I think it is particularly appropriate today in these surroundings and in the shadow of Sir Harrie Massey's death to underline that British and Commonwealth scientists were among the first to realize the potential of space. The British space industry first became established in the early 1960s when it was commissioned to build the Ariel series of science satellites for the forerunners of the Science and Engineering Research Council. The very genesis of the space industry we now have was in our space science programmes and particularly today, and in rememberance of him and his role in organizing this present meeting, I would like to pay my own tribute to Sir Harrie Massey, F.R.S., and to his very great and pioneering achievements in space science. Through his enthusiasm and ideas Sir Harrie inspired a whole generation of space scientists worldwide.

Perhaps because of this early acceptance of space, Britain was also quick to appreciate the potential of satellites for communications; its first links were set up in 1966. It was, after all, an Englishman, Arthur C. Clarke, who first pointed out the potential of the geosynchronous orbit to provide worldwide high quality communications. The U.K. went on to establish its space credentials by being the first Western country after the U.S.A. to design, build and operate its own geostationary communications satellite and we learnt (and then unlearnt!) how to make launch vehicles.

It has been said that this is the decade in which space exploitation has become more important than space exploration. I think it can be rewarding to look back at the key forces in this area before we look forward (the main point of this meeting).

First, I think that it is because space activities have now become an area of mature engineering. It might perhaps be a point for discussion by those following me on whether the time needed to achieve safe, reliable and economic access to space was underestimated just as, early on, remote sensing was oversold. We have had the tremendous achievements of the Apollo programme and the Voyager missions to the outer planets, but both in the U.S.A. and in Europe the development of commercial space transportation systems has taken longer and proved more costly than originally envisaged. Even so, in both places we are now having to consider whether the systems we have in development will have the capacity, economy, flexibility and adaptability to meet our needs into the 1990s. So perhaps a first conclusion is that our first 25 years in space operations have been experimental and exploratory and we still have some way to go.

This points too to a challenge: that future space systems will need to be increasingly evolutionary in their capabilities as new requirements and possibilities emerge.

Second, I think we need to recognize the critical importance of getting to the operational stage with space systems. It is at this point that private investment can begin to take place and commercial benefit result. Because it is only when users can depend on the availability of a particular space or satellite service today, next week and next year that the proper conditions for investment in the systems and reliance on their services can exist. We can see this clearly in communication satellites, where not until we had the assurance of an intergovernmental body, Intelsat, plus continuity of satellites and services could worldwide satellite communications really develop. The same pattern of commitment is possible once there is an adequate promise of operational continuity and service is seen in the reliance by the weather services now put on satellite observations and the way they have taken this routinely into their forecasting machinery. Now we can look back and realize that this was the route that has given us world-wide television coverage and enabled us to have near universal international high quality telephony on a direct-dialling basis.

My third observation may seem initially obvious, that is that space technology does not stand still and that it is necessary to base policy and planning on a lively awareness of this. The problem with space is that the advances have been so rapid and radical that what seems inconceivable becomes commonplace within a fairly short time.

Of course the sums involved in research and development are rather daunting, but by working cooperatively through the European Space Agency, Europe has been able to establish considerable competence and capability in space technology and to keep ahead of new requirements. A good example of such cooperation is in the development of a new large satellite, Olympus, which will be ready to meet the new market demands of the 1990s and in which U.K. industry has a leading role.

With the Olympus satellite project we had in fact very much market-led objectives, but I suspect that with some of the other radical new possibilities that arise in space, you have sometimes, as Mr Beggs of N.A.S.A. has commented, to be prepared to go out and do things without knowing what the results are going to be, otherwise you will never make progress. We are perhaps over tactful in seeking to gloss over a problem that has recurred repeatedly in the past in developing new satellite applications, that the prospective customers say that they have grown and prospered without the new capabilities and services space has to offer.

A related facet evident from our past attempts to provide a regulatory environment for satellite services and on which I would welcome views is that we have to be alive to continuing bounds in technological capability outdating principles that had previously been thought to be fundamental. In the early days it no doubt seemed self-evident that the role of satellites was intercontinental communications linking national telephone systems by using very large dishes.

Perhaps we needed actually to see Armstrong and Aldrin taking their first steps on the Moon to realize that satellite communications had in truth released television from being a local area service, and to mass the market forces to repeal early communication legislation in the U.S.A., which did not foresee and frustrated the domestic use of satellites, and to achieve the great liberalizing regulatory breakthroughs there of the 1970s. I think for the same reason we will have to look carefully through the 1980s at the institutional arrangements (Intelsat, Inmarsat, and soon Eutelsat) that we have established to organize services and share the cost of space facilities. These bodies have been extraordinarily valuable in developing effective and reliable worldwide services and the oldest of them, Intelsat, has shown the importance of a dynamic approach to its mandate and the changing environment of telecommunications. We can learn from this for the future in considering proposed developments in space applications from whatever source, whether institutional or private. The ultimate goal is to enhance the exploitation of space services and to offer a wider choice to the end customer. There is a key economic point here in that, especially in the space field, the earnings to be made from providing services based on using the space systems will often vastly exceed the earnings of the manufacturers of the space hardware while requiring a much smaller capital base.

I hope too that we will always have the courage to learn from missed hopes. We have just had the first flight of Europe's Spacelab. I applaud the courage, skill and endeavour of the crew and all who have contributed to its successful first flight. But we need also to be concerned at the patience of those university and other experimenters who have waited since 1978 to fly their ideas and to whom the prospective future costs of using Spacelab in a continuing way are very daunting. *Nature* (**305**, 655 (1983)), in a recent provocative editorial has called Europe's Spacelab a monument to the past; neither an outright waste of money nor a potential platform for better things but a reminder that software (in the sense of forethought) is more important than developing superlative hardware. Time will judge, with perhaps the dominant factor being how we come to see the importance of space processing.

Looking to the future I would urge a focus on how our space industries may develop up to the turn of the century and on what will be the key technologies and commercial opportunities that will shape their growth. Will it be more communications, sound broadcasting by satellite as well as d.b.s., or space processing or remote sensing, or navigation and search-and-rescue systems?

To encourage you, I think the evidence is that the markets are now materializing outside North America and that they will be large ones. Europe now, collectively, has a large market base, industrial know-how, a rapidly developing need for satellite-based services, a launch capability and sizeable influence in international satellite organizations. For the U.K. I can give you a clear message: my central goal is now to bring space to profit and to pursue space policies in which commercial objectives are prominent.

Let me return now to my starting point. We are fortunate to be living at a time when a totally new world resource has suddenly become available. It is as though the first heavier-than-air flight has just been made. None of us know where it will lead, but we can be sure that even those with the clearest vision can today only see a very small fraction of the ultimate potential.

Can I conclude by inviting a focus on two inter-related issues. How to achieve an effective mobilization, area by area, of the resources of those on the space side, who can foresee promising commercial possibilities with those of the prospective customers, i.e. how to stimulate the latter to plan and make a commitment to participate. Also, I should welcome a focus on the concerns and obstacles that may deter private sector, and indeed public sector, commitment to business opportunities in space.

Phil. Trans. R. Soc. Lond. A **312**, 9–26 (1984)

Printed in Great Britain

Overview of the industrialization of space

By G. K. C. Pardoe

*General Technology Systems Limited, Forge House, 20 Market Place,
Brentford, Middlesex TW8 8EQ, U.K.*

[Plate 1]

This overview paper will set the scene by reviewing the totality of the framework of commercial opportunity. Several mainstream areas such as communications, Earth observations, navigation and the use of space stations will be dealt with in detailed papers later, and this paper will seek only to put their timescale and prospects into perspective. Other areas that have not yet matured to major levels of activity in the process of industrialization will be discussed, to provide a balanced viewpoint. The inter-relations of technology with finance, industrial relations, national and international policies will be reviewed in the framework of the complex nature of the development and exploitation of space projects, particularly highlighting those aspects that relate to the developing countries of the world, who are already deeply involved, and will become increasingly so, in the space business.

The manner in which the world community organizes itself in space activities will have a profound effect on the level of benefits we can gain in the future from our technical achievements. Some of the options will be described and the implications considered of how we move further into the industrial future in space.

1. Introduction

A century or so ago, the countries we now refer to as 'industrialized' displayed an extraordinary degree of innovation in creating, at a remarkable rate, a multitude of factories and service industries. The process accelerated into the twentieth century with new forms of transportation, communication, energy generation and distribution and so on. The rate of progress of the industrial revolution was accelerated further by the stimulae of the two major global wars and now, indeed since the the middle of this century, the process of industrialization has spread with wide implications to the so-called 'developing' countries of the world.

In October 1957, a new era began in the evolution of mankind, when the domain of space was added to the others as one where new commercial and industrial processes could flourish. It is fortunate that space developments are clean compared with the smoke polluted developments of the last century, and are certainly dramatic and exciting. The whole effect has been visibly enhanced by the remarkable advances in global communications made possible by the use of space.

We have today a world community that addresses the equipment supply and operations in the medium of space to generate business of many billions of pounds a year, much of it as revenue from the sale of services. Projected business levels in future years are many times that amount as the whole process takes root even further, and widens its effect worldwide.

The objective I had in mind for this meeting was to pause and take stock of where we have reached in this process, and more particularly to explore where we can, and should, be going in the future. My task in presenting this overview is to put the overall scene into focus, particularly for the non-aerospace related members of the business and industrial community, to assist those who have not so far, perhaps, been deeply involved in the subject, to see some of the linkages and relations between the many sectors of activity and the growth potential in general.

In a subject where the applications of space technology are widening dramatically, a two-day meeting is too short to give full opportunity for certain projects to be presented as separate papers, and therefore I also propose to touch upon a few such projects to ensure that a balanced overview of what is happening in this field is available.

There is of course a clear implication in the use of the word 'industrialization' that commercial benefit is involved, and that is indeed both true and desirable, bearing in mind that the economy of any country depends on the profitable operation of its industry in both the manufacturing and service sectors. The redeployment and investment of profits so earned into the continued improvement and expansion of its industries is for the good of all concerned.

In space we find two distinct categories of operations: one starting and continuing in the field of science wherein immediate commercial return is not appropriate; the other being where technology has emerged already in its application to practical uses, which involves the sale of equipment and services and results in a corresponding revenue from such operations. We are now well into this process already, and it is timely to consolidate our position and drive forward with clear vision towards the right objectives for the future.

2. PRESENT STATUS OF EVOLUTION

Space system elements

It would seem appropriate to start by reminding ourselves just where we have reached in the evolution of practical uses of technology in space. Table 1 simply lists the full range of subject matter that we must include now in our review of this subject. This range of elements involves either the provision of the equipment itself or the sale of services via that equipment or in support of that equipment, or the first together with one of the last two. There are already

TABLE 1. INDUSTRIAL AND COMMERCIAL ELEMENTS IN SPACE:
A LIST OF EQUIPMENT AND SERVICES

space transportation vehicles
    reusable
    non-reusable
communication satellites (space segment)
    point-to-point
        global
        regional
        national
    mobile
        maritime
        aeronautical
        terrestrial
    distribution
        commercial
        entertainment
        education
    direct broadcast

navigation satellites
Earth observation satellites
ground segment (all categories above)
    operational stations
    telemetry, tracking and command stations
    supporting hardware
    supporting software
manned space stations
    recoverable
        Eureca
        Spacelab
    fixed stations
finance sector

major categories in which the business can be grouped and a variety of subdivisions, particularly in the field of communication satellites, and a widening range of activity within the ground segment of all user categories with a massive potential for expansion of business levels in the future.

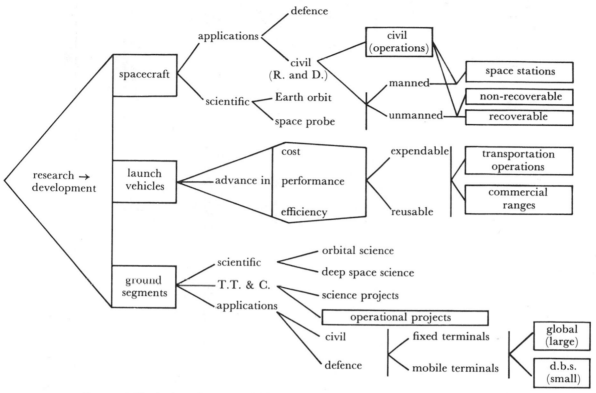

FIGURE 1. Evolution of space activities. Activities in rectangular boxes earn revenue.

Against the simple listing in table 1 it is appropriate to consider, in figure 1 the stages that we have already passed through in reaching the present situation. Table 2 highlights a few of the firsts in terms of technology applications and then operational systems and we come to the present when we have in place various types of global, regional and national communication satellite systems covering both terrestrial as well as maritime mobile facilities; these, of course, will be discussed in much greater depth in other papers. One or two aspects of communications satellites need, however, to be mentioned at this stage, and figure 2 shows the growth in this field; one specific point to note is that there is the curious situation of the missing *aeronautical* communication satellite systems. Starting in the early 1960s, various systems have been conceived whereby satellites would provide air-to-air and air-to-ground communication facilities, particularly over long-haul flights in difficult ocean areas, and even several calls for tender from organizations have taken place with a commensurate amount of wasted industrial effort deployed in submitting bids. So far all such attempts to launch a system have proved abortive and although at this time a modest level of effort continues in monitoring appropriate requirements and planning possible developments, it is a curious fact that as yet no such operational system exists for civil aviation activities. Of course one is aware of the arguments of the airlines

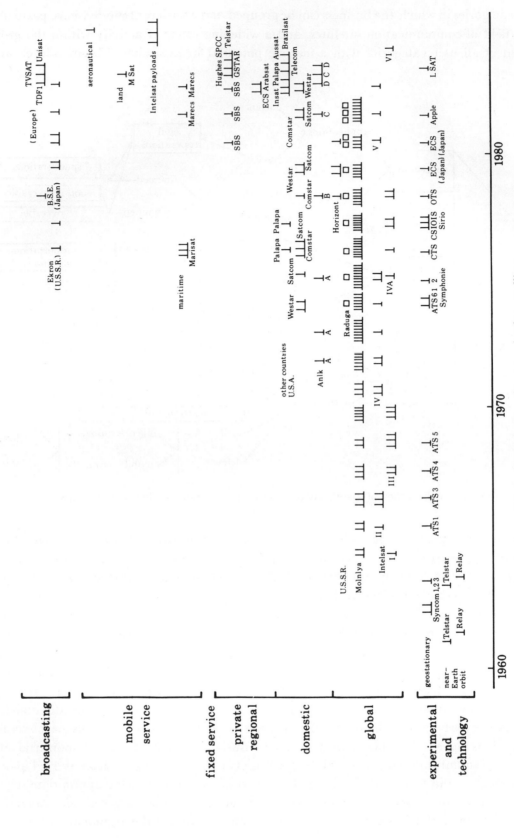

FIGURE 2. Evolution of communication satellites.

TABLE 2. THE STAGES OF INDUSTRIALIZATION

event	year	project
international operations		Space station
	↑ 1990	
commercial materials processing		materials processing factory
inter-satellite communications	1983	tracking & data relay
		Spacelab modular pallets
commercial transportation		manned laboratory
multi-role transportation (scientific, industrial and military)	1980	STS-2 second use of Shuttle STS-1 reusable orbital Shuttle
multi-purpose systems		Marisat maritime communications Skylab materials processing
consolidation		ERTS-1 Landsat images and data
space stations		MARS-2 Mars landing Salyut prototype space station
	1970	
development and exploitation of advanced technologies		Apollo II manned lunar landing
		Venera-4 Venus descent and impact
revenues and commercial operations international commercial collaboration (Intelsat formed)		Earlybird and Intelsat I
global operations		Telstar commercial sponsored communications satellite
practical applications		⌈ Discoverer-13 military reconnaissance
		⎪ Transit navigation satellite
	1960	⎨ Tiros weather satellite
technology demonstration		⎪ Courier Experimental
scientific exploration		⎩ Echo communications
	1957	Sputnik artificial satellite

and those who have been against such systems in past years, namely that the economics must be seen to be supportable and the systems really required before they are deployed. But it does seem that it can only be a matter of time before these criteria are met. It may be not too controversial to suggest that before the end of this decade, a serious and successful attempt will be made to deploy a system of satellites to provide what appears to be an interesting and useful operational service, while bearing in mind the added facility of position surveillance, which is associated with the value of such systems. The aeronautical scene is certainly one that will be watched with interest in this industrialization process.

Education through satellites

Another aspect of satellite communication not dealt with in depth by a separate paper is that concerning its educational services. The world community watched with interest the successful conduct of the SITE experiment in India with the ATS 6 satellite (and indeed the use of that satellite in other parts of the world) to explore the potential use of satellites as a system for the provision of educational services, particularly to developing countries, although with interesting applications and implications for remote areas of developed societies. The success of the SITE experiments is apparent when one now observes that 15 000 to 20 000 villages in India are being equipped with operational read-out terminals, and plans are well in hand for many thousands more to be so equipped to bring the communities within that subcontinent into closer communication with educational services provided by the Nation.

There is a point I would like to make arising out of a paper I prepared some years ago on the subject of operational educational systems in such countries; one of the interesting conclusions from that work was the importance, and indeed the difficulty, of maintaining and servicing a widely distributed system of ground stations in such circumstances. The problem, which is already significant and difficult with several *tens* of thousands of stations in a developing country, becomes quite acute when a system may incorporate a *hundred thousand* or more terminals. Bearing in mind that the eligible number of villages in India that *could* well use such educational facilities is some 520 000, then the subject is worthy of careful consideration. The problem arises not only in the careful specification and design of the ground equipment to operate for as long a period as possible between *regular* maintenance and servicing, but also in the very large numbers of qualified engineers and the associated logistic problems and costs involved not only with this routine maintenance but by the taskforce availability to service the stations that fail at some intermediate time *between* routine visits. This is a separate subject in its own right, but I mention it here because of two aspects. The first is the interesting implications between such dedicated system ground elements and the presence or absence of the domestic service infrastructure, which is firmly in place in developed countries because of the domestic requirements for entertainment or household activities involving electrical or electronic equipment. The second point is the very substantial marketplace for ground equipment, much of which can, and should, be met by local industry; the ground equipment in a typical fully operational system for a country accounts for some 80 to 85 % of the total system value.

Mobile systems

The mobile communication operations on land, such as the Canadian initiative with M-Sat (or Mobile Satellite), are worthy of note. They form an activity that, it would seem, is attracting increasing commercial interest at this time and provides yet another diversity of equipment and revenue-earning opportunity, which must be seen in a healthy context in relation to this whole industrial sector. The other aspects of communication satellites will be dealt with in separate papers.

Navigation satellites

The navigation satellite system is one that has seen considerable success so far and a widening marketplace for both terrestrial and ship based equipment; again to be dealt with comprehensively in other pages.

Earth observation

Earth observation activity is something that I would certainly wish to emphasize in this paper; we shall of course be hearing later of the important work that has opened the way to so much industrial opportunity in both the Earth and space segments and the fascination that emerges from the realization that many users in the world of the information derived from remote-sensing platforms in space, do not as yet realize they are potential users, particularly in developing countries. Table 3 shows simply the main categories of uses. Users divide into two main categories, namely commercial and non-commercial with defence. The situation of users is a problem that faces the world community at present, since the fragmentation and

TABLE 3. USE SECTORS

use sector	state natural	man made
atmosphere	meteorological	pollution
land	resource cartography disaster	land use
water (inland)	resource disaster	pollution irrigation
water (ocean)	resource disaster	pollution (coastal)

remoteness of this marketplace is certainly one of the factors that has led to the slowness so far in the expansion of the commercial uses of this important application of space technology. While in no way wishing to pre-empt the material in other, more detailed papers, I would just mention (to keep the subject in balance) that one observes with interest, certain countries in Europe on the other side of the Iron Curtain (such as Hungary, the scene of the International Astronautical Federation Congress this year), where excellent work is also being done in the use of satellite remotely sensed imagery in relation to aspects of the country's developments.

The other feature to be mentioned is that developing countries are showing a degree of attention to the organizational problems as typified by the creation of the African Remote Sensing Council. We clearly have a long way to go in the expansion of the operational use of Earth observation systems. We must recognize, however, the large humanitarian sector of users, together with the more direct sector of commercial operation: the latter, however, must, by the end of the century, be one of the major areas to be affected by the industrialization of space.

Meteorology

There is no specific paper on meteorological satellite activities elsewhere in this meeting, not in any way owing to its lack of importance, but owing more to the fact that the services have been, and must inevitably continue, to be mainly provided by international or national official organizations. The World Meteorological Organizations, The United Nations, and the various National Meteorological Services (of which that of the United Kingdom has been one of the most active and successful) have created a network of Earth observation communication, interrogation read-out and dissemination systems for meteorological purposes, which have advanced, to a very significant degree, the knowledge of the world's weather and its ability to be forecasted. While the instrinsic value of the information obtained by meteorological satellites is immense in *equivalent* financial terms, it is difficult to see how other than a small amount may emerge (at least in the near future) in terms of directly saleable commercial revenue. It is, however, appropriate to underline the immense importance of this field of application and its close relation to the whole-Earth observation activities that we deal with elsewhere in the meeting.

Space transportation and space stations

The process of injecting our satellites into orbit and, more recently, the process of recovering them, emphasizes that *commercial* transportation in space is indeed with us. It is interesting to note that the U.S. Government has recently named that U.S. Transportation Department as the leading licensing agency for expendable launch vehicles; 'America's newest transportation industry' as Secretary Dole referred to it. Other Government Departments will be involved with labour, safety, marketing and export controls. It is also interesting to note the reference to the development of commercial launch ranges in this respect.

We shall be hearing more in this meeting about the use of reuseable vehicles as well as non-reuseable ones, but it is to be noted that there are groups in the United States and in Europe actively pursuing novel and (they claim) inexpensive methods of providing launch services for satellites by new expendable vehicles. In more orthodox ways, other countries, such as India and China, are developing their own launch vehicles with varying degrees of assistance in the early days from other countries: Japan of course has its family of vehicles that were initially built under license from the United States, but encompassing more home-based technology as time progresses. We must note the reports of reuseable vehicles now being developed in the U.S.S.R. and reports of countries in the Western World taking an interest in the later provision of reuseable or recoverable services in this respect. The subject of space transportation is indeed one of considerable commercial opportunity.

Other papers will discuss developments in space stations; these will highlight the important opportunities of *using* the environment of a space station, and its microgravity field in particular, to great advantage in material processing. It would be hoped that a meeting such as this, held in a few years time, would be able to include separate papers of actual commercial experience, such as the electrophoretic separation work, which appears to be so well ahead as a result of commercial initiatives being taken in the United States already. This work should be watched with great care, and surely any country with industrial interests that may be assisted by the use of a space station environment should be addressing diligently and urgently the question of how to ensure the accessibility of its industries to those facilities in the next few years as we move from feasibility and early development through to prototype operations and finally to actual operational processing capabilities. The commercial use of space stations is a subject little addressed in certain countries so far and yet one that shows signs of immense importance in the future.

Other industrial opportunities, such as the proposals for the LEASECRAFT concept, Spacelab, SPAS, and so on, wherein a satellite platform would be available for leasing to customers to plan their own experiments, alerts the community to the wide scope for commercial initiatives in the space environment.

Long-term developments

Now to consider further into the future. It does not seem appropriate to discuss solar power stations in orbit at this meeting, but this will certainly be covered within the more general large space platforms discussed in Mr Franklin's paper. The momentum that has been built up, modest as it has been in recent years towards this concept, has been steadied by the confusion emerging from the contemporary variations with a downward trend in fossil fuel prices; it would seem that some time has yet to elapse before the world may see whether or not there will continue to be a foreseeable trend regarding the cost and availability of hydrocarbon and nuclear energy sources. Clearly, from work done, the concept of space station energy production on a large scale in the next century is there to be considered more carefully.

These and other aspects of the future will ensure that the space community is never lacking for medium and long-term projects of interest, which move towards commercialization as time elapses.

3. EXAMPLES OF A FEW PROJECTS NOT COVERED ELSEWHERE IN THIS SYMPOSIUM

Spacelab

As explained already, one or two projects would seem to be worthy of particular note although not the subject of main papers. The first of these is Spacelab, which represents a highly successful European project within the responsibility of E.S.A. led substantially by German industry with varied levels of support from other countries within the European space arena. There was some difficulty in scheduling this meeting, to know whether Spacelab would or would not have flown by the time it was held and I am pleased to say that we now know that it has done so, and done so very well.

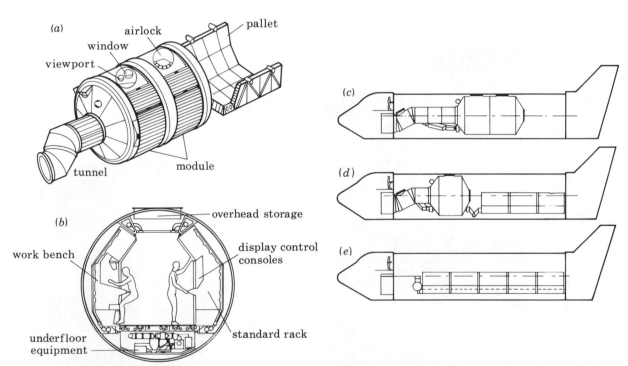

FIGURE 3. Spacelab: (*a*) overall configuration; (*b*) cross section; (*c*)–(*e*) typical Spacelab flight configurations in Orbiter cargo bay, (*c*) large module, (*d*) small module with pallet, (*e*) pallet only.

Figure 3 shows the Spacelab configuration with its flexibility to accommodate variable lengths of pressurized sections to house astronauts for various purposes in exchange for commensurate lengths of the unpressurized pallet to support various other experiments. Figure 4 shows the configuration used in the recent Shuttle flight and figure 5 shows Spacelab ready for installation within the payload bay of the Shuttle vehicle before launch. Figure 6 shows Spacelab's overall configuration. Figure 7 is a view inside the Spacelab with technician specialists at work on their equipment and consoles.

Spacelab represents an exciting and important contribution by Europe to the space inventory of equipment; it provides up to nine days of space station activity in its tethered role within the Shuttle payload bay before its return to Earth. Spacelab represents an important commercial opportunity for experimenters to sponsor a wide range of work that can be accommodated on a

FIGURE 4. Spacelab configuration in Shuttle STS-9.

FIGURE 5. Spacelab ready for installation.

FIGURE 6. Spacelab overall configuration.

FIGURE 7. Internal view of pressurized section of Spacelab.

commercial basis and flown accordingly. A second Spacelab has been ordered by N.A.S.A. in addition to the European one that has just flown, and as such Europe is clearly making a major contribution to manned space work. This experience is most important in providing Europe with first hand awareness of the opportunities and implications of becoming involved in the N.A.S.A.-led space station planning currently underway for the next decade.

Phil. Trans. R. Soc. Lond. A, volume 312

Pardoe, plate 1

MOMS-01-STS-7 RAW DATA ID 1-01-CL-ST7

FIGURE 12. SPAS–MOMS imagery. This is the very first MOMS-1 picture (raw data). It shows the coast of Chile near Africa with the Andes. It is a full-scale picture (138 km × 94 km) with approximately 4720 scan lines with 6912 pixels each. It was imaged during the STS-7 mission on 20 June 1983; the Sun elevation angle was 40.5°. (Data pre-processing and image processing by M.B.B. and R. Haydn (University of Munich).)

Eureca

A further European initiative of importance is the Eureca project and figure 8 shows the general system as planned. Eureca is of course an unmanned spacecraft that will be taken up either in the Shuttle or in a recoverable launch vehicle and injected into orbit, where it will operate untended for long periods. At that time it is capable of being recovered by a

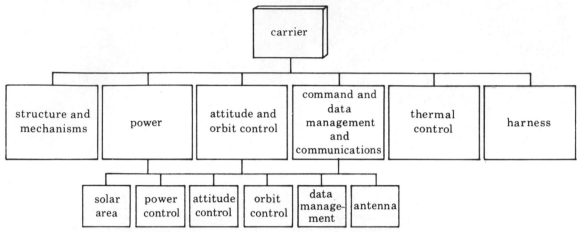

FIGURE 8. Eureca project. Necessary subsystems (middle row) and facilities (bottom row) to operate the retrievable platform and the experiments are shown.

Shuttle and either serviced in orbit or modified with new elements, or brought back to Earth for processing and modification before re-use in a similar mission subsequently. Eureca therefore represents a further important initiative by the European space community. As an E.S.A. programme supported by various E.S.A. countries, again it will be providing immensely important operational opportunity to experimenters from all over the world. It is important that these projects be widely publicized and understood since, although they have been borne of early experimental need, the operational capabilities are significant and it is never too early to move into pre-operational activity.

SPAS

The next project of considerable interest is the initiative taken by M.B.B. in Germany with the development of the SPAS system. SPAS was flown successfully in the STS-7 Shuttle flight of September 1983, and figure 9 shows SPAS within the laypoad bay of the Shuttle in orbit. Figure 10 shows SPAS after it had been taken out of the payload bay by the remote manoeuvring system of the Shuttle, and deployed in orbit. It was moved away from the Shuttle some distance by the self-contained propulsion unit on-board, and in fact took the first photograph of the Shuttle taken from a separate spacecraft close by in orbit, before the SPAS rendezvous and docking with the Shuttle and being recovered by the RMS and relocated in the payload bay before being brought back to Earth.

SPAS contains a variety of interesting equipment as indicated by (figure 11). I would like to point out the MOMS system in particular, which is a self-contained Earth observation package that was used with great success during the STS-7 mission to provide imagery of specific places on Earth, scheduled beneath its flight path, of which figure 12, plate 1, is an example. This

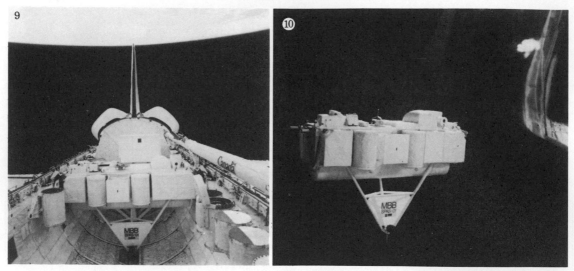

FIGURE 9. SPAS in shuttle bay.   FIGURE 10. SPAS free flying in orbit near Shuttle.

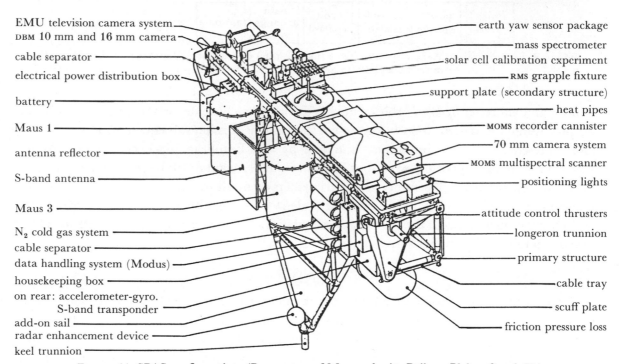

EMU television camera system
DBM 10 mm and 16 mm camera
cable separator
electrical power distribution box
battery
Maus 1
antenna reflector
S-band antenna
Maus 3
N₂ cold gas system
cable separator
data handling system (Modus)
housekeeping box
on rear: accelerometer-gyro.
S-band transponder
add-on sail
radar enhancement device
keel trunnion

earth yaw sensor package
mass spectrometer
solar cell calibration experiment
RMS grapple fixture
support plate (secondary structure)
heat pipes
MOMS recorder cannister
70 mm camera system
MOMS multispectral scanner
positioning lights
attitude control thrusters
longeron trunnion
primary structure
cable tray
scuff plate
friction pressure loss

FIGURE 11. SPAS configuration. (By courtesy of Messerschmitt–Bölkow–Blohm G.m.b.H.)

has been provided by totally electronic means and is imagery to a spatial resolution of 20 m, which is a significant advance on Landsat imagery obtained to date. SPAS–MOMS represents an important operational facility for the future, bearing in mind that an area of observation can be scheduled by commercial contract with the owners of the MOMS system, for a Shuttle flight that may be scheduled with minimum notice because so little (1 m length) of the payload bay of the Shuttle is taken up by the SPAS package, as demonstrated by figure 13. It clearly has the flexibility to be fitted into small sections of the payload bay and is a good indicator and hint for all spacecraft designers of the future who might be interested in having their equipment carried in the Shuttle; namely that a narrow, easily accommodated package is likely to get an earlier flight date than a large cumbersome package taking up more of the payload bay, and will certainly cost less (table 4).

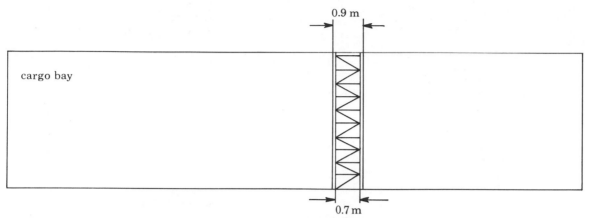

FIGURE 13. To show size of SPAS relative to Shuttle payload bay.

TABLE 4. SPAS LAUNCH COST

	cost
	millions of U.S. dollars
N.A.S.A. nominal shuttle launch cost (1982 value for launches in 1986 and 1987)	70.6
launch cost by N.A.S.A. volumetric charging scheme (for one SPAS element)	$\dfrac{70.6 \times 0.9}{0.75 \times 18} = 4.7$

	mass/kg
mass share covered	$\dfrac{4.7}{70.6} \times 29.7 \times 0.75 = 1480$ (max.)
deduction for structure and payload and subsystems support panels	250
remainder for payload and subsystems and support equipment	1285

The SPAS–MOMS initiative is to be commended and clearly opens the way for initiatives in commercial terms that do not necessarily require levels of expenditure hitherto associated with dedicated flights.

These are a few examples that are indicative of commercial opportunities at present.

4. THE ORGANIZATION OF SPACE

A word now on the important question of how we organize space activity in relation to the industrialization process. Events so far have led to organizations such as N.A.S.A. in the United States, E.S.A. in Europe, and others, as shown in table 5, which also shows some of the procurement agencies in the countries concerned. These Agencies have, with the use of public funds, initiated research and technological applications leading to the design and development of first generations of application satellites. It would seem eminently sensible, and indeed difficult to imagine any other way that such processes could have emerged, bearing in mind the

TABLE 5. ORGANIZATIONS

region or country	space agency	relevant procurement organizations
international and regional	E.S.A.	N.A.T.O. N.I.C.S.
		Intelsat
		Intersputnik
		Inmarsat
		Eutelsat
		A.S.C.O. (Arabsat)
Argentine		A.D.C.S.S
Australia		Aussat
Austria	A.S.A.	
Brazil	I.N.P.E.	S.B.T.S.–Embratel
Canada		Dept of Commun.
		Telesat Canada
China (P.R.)		
France	C.N.E.S.	C.N.E.T.–C.N.R.S.
Germany	D.F.V.L.R.	Bundespost
India	I.S.R.O.	
Indonesia	L.A.P.A.N.	Perumtel
Italy		C.R.A.–C.N.R.
Japan	N.A.S.D.A.	
	I.S.A.S.	
Netherlands	N.I.V.R.	
Sweden		S.S.C.
United Kingdom		D.T.I. M.O.D. S.E.R.C.
		United Satellites
United States	N.A.S.A.	N.O.A.A.
	D.O.D.–D.C.A.	Comsat General
		A.T.&T.
		Hughes Galaxy
		R.C.A.-Satcom
		S.B.S.
		U.S. Sat. Systems
		Western Union
		and others

high capital cost of R. and D. in this particular field. The pattern that is now emerging, however, is that with Government having supported the early R. and D., the industrial sector is then able to identify the commercial opportunities of different elements and assess both investment levels and associated commercial risk involved in picking up such work and deploying it in a normal commercial environment. This is clearly happening successfully in the United States and in Europe with communication satellites of different sorts, and we have had the emergence of Intelsat in the 1960s followed by Inmarsat, in the late 1970s, helping the maritime users. There is now a proliferation in the United States of commercial companies that are privately owned, with different interests in the space communication services field. Such interests have ranged from the provision of total services and the procurement and launch of dedicated satellites for data transmission for entertainment purposes, right through to a large number of companies dealing with specific sectors such as software elements within the total spectrum of industrial opportunity. Naturally the United States, with its predominantly commercial approach to communications, has had a major start on the rest of the world in industrial experience of the problems and opportunities created by such a free market.

The European scene has moved far more slowly I regret, but in recent years the P.T.Ts of Europe have brought the Interim Eutelsat into being, which is now successfully operating its first generation European communications satellite systems for the provision of various services within the Continent of Europe and to surrounding areas. Other regional systems have emerged such as the Arabsat system currently under development, but there are interesting and important examples of national initiatives already, such as that in Indonesia, where for some years. now the national Palapa domestic communications satellite has been in successful operation To date, 132 ground stations have been deployed in the many provinces in Indonesia within the 3200 mile wide archipelago of 13 677 islands, and there clearly exists a massive opportunity in that country for further development of such systems. It is the most creditable demonstration of a developing country using advanced technology to assist in its economic evolution and indeed Indonesia is one example of a country with a strong and professional grasp on the appropriate uses of advanced technology for national development purposes. Similar initiatives are evident within India, which has been mentioned before in relation to education satellites and again a national organization, I.S.R.O., has been set up to look after much of the space work.

It is not appropriate or necessary to catalogue all the various organizations that exist at this time in various countries, but the ones mentioned so far have been so referred to in order to make the point that the process of industrialization is fundamentally aided by governments looking ahead and taking initiatives with public funding in the *early* stimulation and application of scientific and technical endeavour, and then a carefully phased hand-over (or at least joint involvement) to industry as appropriate events dictate. What *is* essential is to provide an adequate opportunity for industry – if it so wishes to invest in space – *to gain access to revenue*, because unless a fair and reasonable coupling is provided, clearly industry cannot be expected to invest without methods becoming available to service its investment in the interest of its shareholders. This lack of willingness of earlier U.K. Government Administrations to allow access to operational revenue within the U.K. has certainly been one of the major inhibiting features of industrial investments in the early 1960s when private initiatives were being sought for this commercial use of space communications facilities.

It must be said, however, that the counterpart is also important to note, namely, that countries where high levels of Government expenditure have been evident in the first few years of the technology development process run the risk that industry becomes used to any easy, captive marketplace and instead of the public funds being used to seed and stimulate a normal independent industrial development, it can be looked at by industry to be a market *in its own right* and consequently weak and ineffectual international marketing and planning takes place in industry; such attitudes do not well equip such industries with experience to address the open competitive markets of the world. It is essential that industry moves in its own right at an early date, otherwise it may enjoy for a limited period the easy government generated levels of business, but these *must* be at risk in the finite length of time that these subsidized projects can be allowed to continue. Unless a lean, hungry and competitive industry has emerged from the early years, it will not penetrate the world markets, compared with those countries that have experienced a tougher and rougher entry into the world marketplace. A quick examination of the non-aligned and third world contracts placed for the systems for communication satellites, for example, which are in place or under development at this time, reveals that the main success has come from companies in the United States, Canada and France.

The final point I wish to make on this question of the organization of space industry is the need for careful planning to identify both the opportunities as well as the problems. Such planning must be made effectively by both governments and by the industries concerned, as well as that available from independent consultants, because unless the most imaginative and well founded programmes are initiated, the projects eminating from them will be more at risk than others that have not been well thought through.

5. International cooperation

Of course the nature of space, both in its size and scope, lends itself to international cooperation and much has taken place already. This has often been between companies and between collaborating agencies of different countries; perhaps the most visible and formal cooperation arrangements have been made within Europe. Of necessity, international European consortia have emerged to bring the technical expertise into a common team and also to assist in the allocation, with some reasonable logic, of contributed public funds of each country back into the industries of those countries. This led to the three main spacecraft consortia in Europe emerging in the 1960s and operating throughout the 1970s, although in recent years some of the national mergers and changes in industry have, of necessity, blurred the shape of the membership of the consortia and so at this time perhaps more *ad hoc* groupings have become apparent in addressing European and world markets.

Space station

Looking to the future, the question of international industrial collaboration is very essential when considering projects such as the space station. Looking back to the post-Apollo era and the extensive studies, discussions and negotiations, on what part Europe might play in that programme, I remember well how frustrating and difficult it was to get agreement on what part Europe would play in that work. Spacelab emerged from those considerations – but not without a considerable amount of wasted effort – going through the possibility that the Tug might be the European contribution and indeed looking into possible collaboration in the Shuttle itself. So looking ahead now to the space station, a degree of realism is essential as to what parts *might* sensibly be eligible for international participation; in looking at this it should not (I suggest) be assumed that access to a space station facility will necessarily be related precisely to whatever funds other countries or continents may make available for space station *equipment* that may be *supplied* from those sources. Several countries have a possible interest in collaboration with the United States and the space station and these so far include Canada, Japan, and Europe. It can be seen that careful thought is necessary to explore just how all of these interests might come together amicably and effectively to lead to good operational facilities in the 1990s.

The nature of the space station is such that it would seem eminently sensible that international cooperation *does* take place and it would appear to be the intention of the United States at this time to encourage it. It behoves the other countries therefore, to look carefully to the opportunities and implications of such cooperation and to move firmly and in a timely manner to secure an appropriate position in this respect. The ultimate factor of course, is the level of user interest. and who will wish to pay for time in these various space station facilities. It is important that potential collaborating countries, companies, and indeed *all* possible users, address this matter as carefully as possible at this time. We are talking, of course, of operational use many years away in the future and much further development and thought about the best way to use the space environment can, and will, emerge in that time.

I would not wish to limit these few words on cooperation simply to space stations, so I note also the benefits in the world marketplace of appropriate industrial teaming to bid for various markets, and note that it would seem that an important criteria will be getting the balance right between long-term commitments with various partners while having flexibility to change teaming arrangements as further needs emerge in other parts of the world.

Developing countries

A particular comment needs to be addressed to the question of developing countries, which represent a very important marketplace in the future for the technologies and equipment, software and expertise in general supplied by the *developed* countries. I would suggest, however, it is essential not to think of this as one-way traffic where developing countries are buying and remaining completely dependent on developed countries, but rather a two-way business where, of course, initial critical parts of systems, particularly in the space segment, will be provided by developed countries. But increasingly the *developing* country's industries will take part in the ground segment and thus permeate steadily but surely into responsibility for the provision of, and operation of, major parts of the system. Bearing in mind the remarks earlier concerning the high percentage of total system values associated with the ground segment, there is scope for all in meeting the operational requirements of the future in the developing countries. The work with which I am involved in Africa and elsewhere at this time, underlines fully the importance of good collaborative arrangements with local experts and local industries as well as the governments, to secure sensible and successful programmes for the future.

6. FINANCIAL IMPLICATIONS

The 'bottom line' of any development programme of course is the question of funding. As I have just discussed, the nature of space as it has evolved has led to a very significant deployment of public funds in the early years, but with a progressive transient take-up by the private sector. Major opportunity therefore exists for the finance houses at all stages of the work, which ranges now through the space transportation processes, the deployment of satellites themselves, and certainly the provision of the ground segment; and this is why we should study carefully the paper by Mr MacArthur on this finance sector.

A point of great importance is the relation between technology and finance in space projects; without the right approach to funding sources by industry, no commercial project will succeed. Without an ability by the finance sector to appreciate the nature of technical opportunity, and risk, associated with the advanced technology of space, projects will not be successful. The interface is vital and operational procurement organizations, such as Eutelsat and Inmarsat, clearly have a great need to recognize this interaction: for example, in the preparation of appropriate cost models showing the parametric variation to be expected due to technical change, the degree of sensitivity of costs to specification or performance alterations, etc. Without wishing to over emphasize the point, these complex relations demonstrate the breadth of disciplines necessary in the industrialization process.

I also wish to emphasize the importance and opportunity in the insurance aspects of equipment particularly related to the space sector. Large projects are involved and correspondingly large sums of money and insurance cover of several hundreds of millions of dollars attracts serious and professional thinking and planning in the determination of premium levels as well as the payment of the insurance cover in the event of any regrettable failure! We have seen evidence of all of these aspects in recent years, and as a consultant concerned with assisting banks and insurance companies in both the planning, as well as the subsequent failure mode assessment of such systems, I am certainly in a position to underline the importance of recognizing the role of finance and its relation with technology in this whole space field. We are, of course, now acquiring an increasing amount of statistical data with regard to launches and satellite performance functions, but the complexity is also increasing, and so the question of insurance is a small but vital expanding sector of the whole process of commercialization and industrialization.

7. CONCLUDING REMARKS

We are already well into the process of the industrialization of space and serious and extensive commercial activities are widely underway in all parts of the world. Space is big business and will continue to be so. Levels of space business are currently measured in several billions of pounds per annum and when one combines the opportunities of *supply* of the equipment as well as the *sale* of services, then we are looking at many tens of billions of pounds of business available to the industries of the world even within the rest of this decade. It is therefore not only exciting and dramatic, but a most serious business sector; and yet it is still considered by many of the non-aerospace or electronics fields of activity as something apart, not to be taken too seriously. I refer, among other things to the implications within the wider use of Earth observation data and also to the users of space stations. Experience has shown that not only is an intense and active campaign necessary to ensure awareness of the opportunities in different sectors of industry outside of aerospace, but there is a great necessity to *follow through* very extensively to consolidate such early interests that may have been seeded, and to assist and advise many companies of how best to take up an undoubtedly interesting and opportune role in this field. We must get the market pull lined up properly as well as the technology push, which is generally very apparent!

This is extremely important in the non-aerospace industries where potentially beneficial developments exist; an example is the pharmaceutical sector, in relation to which I have a list itemizing over 20 selected pharmaceuticals *each* of which sells for over $1 billion per kilogram (the highest is $20 billion per kilogram) and these are being actively considered as candidates for space environment processing, with all the benefits of higher purity and higher yield than can be achieved in Earthbound facilities. When such benefits are seriously in prospect, no one can afford to be superficial in assessing the possible industrial implications.

I hope that these few remarks have therefore set the scene for the rest of the meeting, which will cover key areas in more detail. For my part, I am delighted that so many people have come together at this time to take stock of the current situation and to address the exciting opportunities of the industrialization of space in the future.

Phil. Trans. R. Soc. Lond. A **312**, 27–32 (1984)

Printed in Great Britain

[27]

International communications satellites: global and European regional systems

By A. K. Jefferis

British Telecom International, Holborn Centre, 120 Holborn, London EC1N 2TE, U.K.

The development of the Intelsat network has caused a revolution in international communications for most countries and particularly those in the developing world. The high growth rates in international communications have resulted in a new satellite generation every few years, which use progressively more advanced technology in both the space and Earth segment. However, Intelsat VI may be the end of the trend to ever larger, more complex satellites.

Satellites were unable to compete effectively with Europe's highly developed terrestrial network. However, European Governments took a view not only on the need for cheap communications but also the need to develop a capability in the field of satellite communications that would equip industry for the massive world markets foreseen for the future. The result was ECS, a regional communications Satellite system for Europe that will go into service within a few months and will be used well into the 1990s.

Introduction

Satellite communications must represent the most significant industrialization of space in the first 25 years of space exploration. This can be well illustrated by both the Intelsat global system and the European regional system, and these also serve as good examples of where satellite communications are likely to lead in the next decade or two.

Although prime interest is in the future we cannot predict this without reviewing what has happened in the past.

Intelsat

Intelsat started in 1964 with 11 members and one satellite. It now has 104 members and 17 operational satellites. The organization was formed by an agreement between Governments (the Parties) who meet every two years in the Assembly of Parties. There is also an operating agreement between the telecommunications authorities nominated by the Governments: the Signatories. They meet once a year in the Meeting of Signatories. The practical direction of the system is determined by the Board of Governors, which consists of the 13 largest Signatories plus certain other groups of smaller Signatories.

Signatories contribute to the capital requirements of the system in proportion to their use of it. They then pay for their use of the system on a per-circuit basis at a rate designed to give 14% per annum return on capital and to meet the operating expenses and depreciation. This has resulted in a charge that started at \$32 000 per annum for one half circuit in 1965 but now stands at approximately \$4 700 per annum.

To see what the industrial opportunities will be in the future a brief review will be given of the operational and technical development of the system since 1964.

There are three operational regions, each with at least one operational satellite and a spare one in orbit. The Atlantic Ocean Region covers much of North America, South America,

Africa, Europe and the Middle East. The Indian Ocean Region covers Europe and Africa to Japan and Australia. The Pacific region covers the whole of the Far East and the west coast of North America. Two-satellite operation is introduced when the traffic in a region is too much for one satellite and also when the users feel that they are not prepared to put all of their circuits through a single satellite and a single Earth station aerial. One of the satellites is then designated the primary, to which practically all of the countries in the region operate and which therefore provides full connectivity in the region.

The second satellite is termed the major-path satellite and carries typically half of the mutual traffic of those countries possessing second Earth station aerials. These larger countries therefore split their traffic equally between two satellites and two Earth terminals. In the Atlantic Ocean region there is now also a second major-path satellite used by countries with three aerials, while the Indian Ocean region has a primary and one major-path and the Pacific region a primary satellite only. In addition to these operational satellites each region also has a spare satellite in orbit.

This configuration of the global system has led naturally to an evolution whereby a new generation of satellite is introduced every few years (typically five) and is put into service as the Atlantic Ocean primary and in the spare role. Since satellite lifetime is typically 6–8 years the replaced satellites can be moved to fulfil some of the less demanding roles. Gradually the new generation satellites are introduced into the successively less demanding locations and the predecessor satellites retired at the end of their lifetime. In practice some of them have also served for domestic leased services.

Intelsat services

The bulk of the Intelsat capacity is used for intercontinental telephone-type services, but there is significant use for television transmissions, a service that has now developed to include some full-time international television leases between certain countries in addition to the traditional service of transmissions booked on an hourly basis.

Much of Intelsat's spare capacity is used for domestic services, which are provided on a leased transponder basis. (A transponder is a radio transmission channel through a satellite; it is typically capable of carrying one or two television channels or from 500 to several thousand telephone channels.) Also, Intelsat has recently introduced international business services permitting access via small Earth stations, which may be at or close to the end-user premises where this is appropriate.

Evolution of intelsat satellites

Since its first satellite, Early Bird, otherwise known as Intelsat I, Intelsat has used a further five generations of satellite design, the principal features of which serve to indicate the development.

Early Bird was a spinning satellite built by the Hughes Aircraft Company that provided about 240 telephone circuits through a partially directional aerial, which gave coverage of the North Atlantic. It could permit only two Earth stations to access it at a time and it used only about 50 MHz of the available 500 MHz of bandwidth. Its mass in orbit was 38.6 kg.

Intelsat II was also built by Hughes and was similar to Early Bird in capacity but it had a single multiple access transponder of about 120 MHz bandwidth.

Intelsat III, built by T.R.W., had two transponders filling the available 500 MHz bandwidth and providing some 1500 circuits by using an aerial fully directed at the Earth and giving global coverage.

Intelsat IV, built by Hughes, had 12 narrower bandwidth transponders filling the 500 MHz available bandwidth and provided some 4000 circuits plus a television channel.

Intelsat IVA was similar, but for the first time permitted re-use of the same frequencies on a given satellite to provide approximately 50% more capacity. It used 20 transponders connected via aerials that gave beams shaped to the contours of the continents to permit frequency re-use between East and West. All the satellites up to this time were spinning to achieve stabilization relative to the Earth's axis.

Intelsat V, built by Ford Aerospace, is fully stabilized rather than spin-stabilized and further extends the capacity to some 12000 circuits plus two television channels by fourfold frequency re-use, introducing polarization separation for the first time. It also introduces the new frequency bands at 11 and 14 GHz and it has 27 transponders.

Intelsat VI, now being built by Hughes, has over 40 transponders with an even greater degree of frequency re-use than Intelsat V, but, more significantly, introduces for the first time on-board digital switching that, together with appropriate new techniques in the Earth stations, can raise its capacity to some 30000 circuits plus several television channels.

Over this period of some 20 years the design lifetime of the satellites has increased from 18 months for Early Bird to 10 years for Intelsat VI.

FUTURE FOR INTELSAT

Technically the trend in Intelsat is likely to depart from its past pattern of a bigger and better satellite every few years because it is becoming progressively more difficult to obtain greater capacity from a single orbit location because of the constraint on the available frequency bands. Over the years Intelsat satellites have been one step ahead of other satellites used in domestic and regional systems, but there is now a risk of Intelsat getting too much out of step with most other applications.

Although there are further frequency bands at 20 and 30 GHz these are unlikely to be technically suitable for increasing capacity in Intelsat for at least a decade. Rather, increased capacity is likely to be achieved by new techniques in the Earth segment, particularly those associated with more efficient digital transmission systems.

Intelsat is likely to introduce more than one type of satellite in a given generation to meet the diverging requirements of the primary satellites in the Atlantic and Indian regions, and the various major-path, Pacific region and domestic needs. These other satellites are still likely to be relatively large, at least Intelsat V or Olympus class, and the satellites for the primary roles will need to continue to be at least of Intelsat VI class.

The principal development in satellite design is likely to be the introduction of a greater degree of on-board digital processing, while a significant improvement in system capacity will come from new digital techniques introduced in the Earth segment.

Industrially, these trends should give greater opportunity for industry outside the U.S.A. to participate in Intelsat programmes since the demands of the satellites will not be so far ahead of other systems that will be progressing in parallel.

Intelsat may also have to face changes in respect of competition going beyond the regional systems that are now permitted. In particular there is the threat of competition on transatlantic services, proposed by Orion, and from other systems. In principle, these proposals are contrary to the Intelsat agreements and it would seem unlikely that they will be authorized. If they are, however, they are likely to lead Intelsat changing its principle of uniform charging throughout its network, which will have an adverse economic impact on the many smaller routes that are currently provided. Such a change could also alter the types of satellite that Intelsat requires to provide its services.

Finally, politically there are moves by some third-world countries to change the procedures by which orbit locations are allocated for communication satellites. Any move towards detailed planning of orbit use would have a serious impact on Intelsat's ability to maintain and expand its global services.

THE EUROPEAN SYSTEM

The situation in Europe has in many respects paralleled that in Intelsat, but on a smaller scale. In 1970, E.S.R.O., the predecessor of the European Space Agency (E.S.A.), and the European Conference of Post and Telecommunications authorities (C.E.P.T.) conducted studies of possible operational systems for Europe, but the results showed that no economic case could be justified at the time. As a result, E.S.A. went ahead with the O.T.S. satellite, which was launched in 1978 and was a pre-cursor to a possible operational system.

In the mid-1970s further C.E.P.T. studies led to the conclusion that an operational system starting in the early 1980s would be justified if the Governments continued their funding to cover the provision of the initial satellites and the costs of their development. This they agreed to do to give European industry a foothold in expanding world markets. As a result, a decision was taken to proceed with the E.C.S. system consisting of four satellites plus a ground spare designed to provide two satellites in orbit throughout a 10 year mission.

In 1977 Eutelsat was formed, mirroring Intelsat in many respects, to run the E.C.S. system in return for annual payments to E.S.A. The E.C.S. satellite has 12 transponders of which 9 can be operated simultaneously, the other three serving as spares. From the second flight model onwards there are two further transponders provided for business services. The satellites have three spot beams in addition to European coverage and operate entirely in the 11 and 14 GHz frequency bands.

There will be typically one Earth station per country for the main mission, which will use digital transmission for telephony and frequency modulation for the television. Business services Earth terminals will use small aerials of 3.5 or 5 m diameter. Eutelsat will also be using one transponder in the Telecom 1 satellite for a certain category of business applications. A new dimension, not originally foreseen for Eutelsat's activities, is the leasing of transponders for television distribution to cable television networks for which otherwise spare transponders will be used.

FUTURE OF EUROPEAN SYSTEM

By the end of 1985 the European system may need a third satellite in orbit to cater for the increased demand for leased television transponders. Launching the third satellite will bring forward the date when the next generation will be required.

One possibility for the next generation is the so-called E.C.S.-A, which would be a modest development of the initial E.C.S. satellites sufficient to carry the main mission telephony an television transmissions until the mid-1990s. The carriage of the increased level of traffic will be greatly helped if a technique to reduce the transmission rate for digital telephony is introduced into the Earth stations. This should be possible by 1990 and will avoid undue complexity in the satellites. The satellites will also have to carry the business services provided on the first generation, preferably including the component that is initially being carried on the Telecom 1 satellite. Provision for television leases, both fully spared and preemptible, is likely to require the continuation of a configuration with three satellites in orbit.

In the longer term, additional capacity could be provided by the introduction of 20 and 30 GHz frequency bands, frequency re-use other than by means of polarization, i.e. by multiple small spot beams, and may possibly involve the use of on-board digital switching and processing.

Since the economics of satellite communications in Europe other than for television distribution are not sufficiently favourable compared with terrestrial systems, it seems unlikely that Eutelsat will face competition from other satellites within the next decade, except for competition for its leased transponders from Intelsat or national systems. This economic picture is likely to force Eutelsat to follow a relatively modest development of its services, concentrating on well tried techniques and operating its satellites as close to saturation as possible.

Politically, there may be some complications introduced into the way Eutelsat operates if more countries follow the U.K. lead of introducing competition nationally. Also, procurement policy will not necessarily reserve this market for the European space industry alone.

As a general conclusion, therefore, there will continue to be opportunities for the European space industry in maintaining and expanding the European regional system, but the development is not likely to be as dramatic as it has been in Intelsat.

Discussion

F. MILES (*Independent Television News, London, U.K.*). Mr Jefferis has referred to the remarkable lowering of the cost of a 'half circuit' when using communications satellites, in comparing 1964 charges with those of today. But why is it that when using satellites for television transmissions across the Atlantic, British Telecom charge around twice as much for their 'half circuit', that is, bringing the signal *down*, as an American company will charge to put it *up* to the satellite?

A. K. JEFFERIS. The basis of charging for television transmissions frequently differs from one operator to another. For example, in the U.K. the cost covers the transmission from the international Earth station to the customer's premises within the U.K., but this is not always so in the U.S.A. Also, Comsat frequently make charges without taking account of the proper full allocation of the costs of their Earth station, i.e. they charge the fully allocated costs to the main telephony services but carry television on a marginal basis. I am not able to comment on whether your claim that the B.T. charge is around twice as much as that charged by an American company. I will be happy to look into this and give Mr Miles appropriate information after the meeting.

(After the meeting, Mr Jefferis established that in fact the B.T. charges are substantially *below* those charged by the American companies that offer equivalent service.)

D. O. FRASER (*British Aerospace p.l.c., Dynamics Group, Space and Communications Division, Bristol, U.K.*). In addition to the return on capital from Intelsat, has Intelsat use been profitable to the member countries?

A. K. JEFFERIS. Intelsat itself does not make profit for its members. As I explained, the Intelsat charges cover operating costs, capital depreciation and enough to pay 14 % per annum return on the capital invested by the members. However, in using the facilities provided by Intelsat as part of their international networks, most of the members do of course make substantial profit from their international telecommunications activities. For British Telecom, this can be seen quite readily from the published Report and Accounts.

I. NICOLSON (*Hatfield Polytechnic Observatory, Bayfordbury, Hertfordshire, U.K.*). Could Mr Jefferis indicate the approximate annual turnover in terms of revenue and capital of Intelsat at present?

A. K. JEFFERIS. The 1983 figure for revenue, which covers operating costs, return on capital and an allowance for depreciation, is about $400 million.

Phil. Trans. R. Soc. Lond. A **312**, 33–38 (1984)   [33]
Printed in Great Britain

African satellite communication systems and their implications

By Y. Demerliac

Eurospace, 16 Bis, Avenue Bosquet, 75007, Paris, France

Eurospace, founded in 1961, is the Organization of the European Space Industry. Its members are 80 major industrial companies, banks and operators from 13 countries in Western Europe. The Association is the mouthpiece of industry to the European Space Agency (E.S.A.), in particular with regard to future European space programmes. In the course of its other promotional activities it has applied and sustained successful efforts to promote a regional communication satellite system serving the African continent.

Introduction

The size of the African continent and, in general, the poor situation of its communication facilities almost inevitably call for solutions that use satellites and, for a long time, many proposals for such solutions have been made. This paper will deal only with the so-called Afsat project promoted and studied by Eurospace and which, at this stage, appears to be the most promising approach to a regional African communication system.

Initially, the Afsat concept arose from the conjunction of thoughts of the U.A.P.T. (African Post and Telecommunications Union), a group of 12 French speaking countries and Eurospace. This concept is based on a triangular arrangement that involves U.A.P.T. in the role of the African customer, with the European Development Fund of the E.E.C. providing the finance and the suppliers. Eurospace and its group of consultants (E.S.A., Detecon, General Technology Systems, Satel-Conseil, Safritec and Telespazio) have so far acted as study 'suppliers'. When the studies are completed industry will take over from the consultants and will supply the required hardware and services.

It is worth noting that from the outset U.A.P.T. undertook to widen the scope of its approach by involving English speaking countries. This led to the setting up of an *ad hoc* Afsat Committee to which seven English speaking countries (including Kenya and Nigeria), Cameroon and Zaïre participated in addition to the 12 U.A.P.T. member countries. Two studies made in accordance with this scheme are now available: a study of the telecommunication, radio and television requirements of the countries participating in the Afsat Committee, and a parametric feasibility study covering the whole continent. The latter was delivered to, and approved by, the customer in November 1983. These studies will be now summarized. The current problems regarding the next steps will then be reviewed and some conclusions drawn.

Current situation and needs of communication in Africa

Except for the links provided by Intelsat, and submarine cables between certain African capitals and between them and other continents, current communications in Africa are far from being satisfactory for both telecommunications proper (telephone, telex, data) and for radio and television.

For telecommunications proper, in spite of the efforts undertaken to implement the Panaftel network, a system of microwave links aimed at providing communication between the countries and at creating a communication 'axis' inside each of them, the situation remains difficult. It is almost impossible for many African capitals to call each other on the telephone except through Paris or London! Inside the various countries, the density of direct exchange lines (d.e.l.) is very small (seven d.e.l. for 10 000 inhabitants in Nigeria, three for 1000 in Senegal, seven for 1000 in Zambia). Moreover, density figures are misleading because they are country-wide averages. One must realize that, in general, 80 % of the telephone sets are in the capitals, the rest in the large cities, almost none are in the rural areas, whereas the latter represent 80–90 % of the population. However alarming these figures may seem they do not fully reflect the very poor situation, because the rate of availability of existing lines is much less than one owing to the line saturation in the large cities and the transmission problems in the inter-urban links.

For television, in general, only the capitals and adjacent areas are provided with service. The radio coverage is better but needs considerable improvement, particularly F.M. broadcasts. The assessment of the communication needs in this context is a challenging exercise. To aim for current European standards – say one d.e.l. line for three to five inhabitants, with three television channels covering the full territory of each country – would be too ambitious. Thus, for telecommunications proper, the Afsat mission definition for the period 1988–1997 took into consideration both the channel requirements assessed by each country and their economic capabilities. This resulted in a mission requirement of 8000 F.D.M. channels, 17 000 single channel per carrier (s.c.p.c.) channels, six television channels, which, by using the system presented below, requires a satellite capacity of 20 to 31 active transponders in 1997 (table 1).

TABLE 1. SUMMARY OF THE TOTAL THEORETICAL SATELLITE
CAPACITY ENVISAGED IN 1997

service	number of transponders	
	minimum	maximum
interstate telephony	3	3
inter-urban telephony	11	18
rural telephony	2	4
interstate exchanges of radio and television	1	1
national television	3	5
national radio	0.10	0.10
total	20.10	31.10

Interstate telecommunication and television services are provided to the whole continent, whereas domestic similar services will be available only to countries south of the Sahara. Telecommunication quality standards are those in C.C.I.R.–C.C.I.T.T. relevant recommendations. For television distribution, current Intelsat standards were adopted.

FEASIBILITY STUDY OF THE SYSTEM

To fulfil the mission described, the study proposes an optimized system comprising a space and a ground segment. Four space segment configurations ranging from 20 to 36 transponders were examined. The reference configuration used later on in economic computations is based on a high power 24 transponder active satellite of the Arabsat or Eurostar class. One idle back-up satellite in orbit and one spare satellite on the ground, ready for launch, are foreseen so as to provide guaranteed services. The launcher will be either Ariane IV or the U.S. Shuttle.

The payload characteristics can be summarized very briefly.

(*a*) Four 'global' beam transponders are allocated to interstate services, each of them covers the whole continent (effective isotropic radiated power (e.i.r.p.) 31 dBW, sensitivity -10 dB K^{-1}).

(*b*) Twenty 'spot' beam transponders cover four to six domestic service zones (e.i.r.p. 37 dBW, sensitivity -3 dB K^{-1}). When using 4.5 m diameter ground stations, the telephone capacity of each transponder is 1333 channels with a companding factor of 13 dB. Alternatively two television channels can be transmitted per transponder in good quality conditions. (One of the configurations offers six higher power dedicated television transponders enabling two televison programmes to be transmitted per transponder with reception by ground stations of less than 3 m diameter.)

For the ground segment, three station classes were adopted.

(i)   Class 1 are master stations to be located in the capitals. Typically their diameter is 11 m. They include a tracking system and can handle from 96 to 600 F.D.M. channels and 60 to 200 s.c.p.c. channels for interstate, inter-urban and rural telecommunication purposes. They are capable of transmitting and receiving television signals.

(ii)   Class 2 stations will be located in large towns. Their diameter is 8 m and they require no tracking system. Their telecommunication capacity is from 24 to 96 F.D.M. channels and 12 to 60 s.c.p.c. channels. They can receive (but not transmit) television, with a rebroadcasting capability in very good conditions. No tracking is required.

(iii)   Class 3A stations are typically 4.5 m in diameter and intended for small towns or large villages. Their telephone capacity ranges from 1 to 12 s.c.p.c. channels, they can receive television with an acceptable rebroadcasting capability. No tracking is required. Class 3B stations are 3 or 3.5 m in diameter and will be mainly located in small villages or company premises. Their capacity ranges from 1 to 5 s.c.p.c. channels. They can receive television with no rebroadcasting capability. No tracking is required. Self-contained power supply units with solar cells are proposed to power the class 3A and 3B rural stations where no electrical network is available.

It is important to note that to optimize a communication satellite system a trade-off is always required. Given certain transponder characteristics large ground stations are complex and expensive in terms of investment cost, but they require less transponder bandwidth and less power transmission per channel than the smaller ones. The Afsat system concept presented is the result of an iterative process that enabled the characteristics of both its space and ground segments to be so determined as to provide an overall combination that is at the same time most economic and best adapted to the African environment.

Economic aspects

A full study of the system's economic viability could not be made because the study budget did not provide for data collection in Africa. So, in particular, income assumptions could be made only for a limited number of countries. However, the investment and operational cost could be assessed precisely for the whole space segment and for the use of Afsat by a reference country (300000 km², 10 million inhabitants in 1990, one master station, 29 inter-urban stations and 90 rural stations to be deployed in the last 4–5 years of the period).

Over 10 years the total investment cost of the Afsat space segment for the whole continent is assessed to be $200 million. The cost of the ground segment over the same period for the reference country amounts to about $26 million. For such a country, the operating cost of the ground segment would be $2 million per year. On this basis the total ground segment investment for all Africa will be between $600 million and 800 million, depending on the number of participating countries.

Based on a maximum transponder leasing cost of $3 million per year, it was found that the total unit cost per domestic channel (i.e. by combining investment and operational costs of both the space and ground segment) would amount to $7200 per year per inter-urban or DAMA rural voice channel, $1 600 000 per year per television channel (marginal cost of distribution of one television programme to 10 ground stations).

Here it must be emphasized that the economic results mentioned can be considered as very conservative. In particular the leasing cost of $3 million per year per transponder is based on a comparatively low use rate of the satellite's capacity (70 % on average over the mission period) and a high internal rate of return (14 %). Moreover, no provision was made for the leasing of capacity, on a preemptible basis, in the back-up satellite.

Finally, the cost of using Afsat by the reference country when its ground segment is fully deployed amounts to less than 0.4 % of the country's projected gross national product at the end of the proposed mission period in 1997, a value that can be considered reasonable.

COMPARISON WITH ALTERNATIVE SOLUTIONS

The technical and economic Afsat data will now be compared with alternative space and terrestrial solutions. For space solutions Intelsat appears to offer the most credible alternative to Afsat and, considering their currently existing and planned programmes, a comparison of Afsat was made with leased domestic transponders in Intelsat VI 'zone' transponders (e.i.r.p. 28 dBW, $G/T = -7$ d BK^{-1}); Intelsat VA 'spot beam' transponders (e.i.r.p. 32.5 dBW, $G/T = -18$ dB K^{-1}).

It was found that if the same Earth stations foreseen for the Afsat mission are retained when an Intelsat VI or Intelsat VA satellite is employed, it is necessary to increase the transmitter power by a factor of 10 to 25 and to sacrifice a half or two-thirds of the transponder capacity. Alternatively, if the diameter of the station is changed (from 4.5 to 8 m for the class 3 stations and from 8 m to 11 m for the class 2 stations) a transponder capacity between 50 % and 100 % of that of an Afsat transponder can be obtained, but this will demand the use of an autotrack system for class 2 stations and the increase of transmitter power by a factor of 3 to 10.

In economic terms it was found that for a guaranteed service, the use of the Intelsat space segment would lead to a total network cost: from 32 % to 105 % higher than Afsat for inter-urban services; from 30 % to 64 % higher for rural services; from 34 % to 101 % higher for television services; from 48 % to 63 % higher for the complete network. For the *non-guaranteed* service, the use of the Intelsat space segment would lead to a total network cost: from 19 % to 48 % higher than the *guaranteed* Afsat inter-urban srvice; from 22 % to 60 % higher than the *guaranteed* Afsat rural service.

So it appears that a *guaranteed* Afsat service is economically more advantageous than a *pre-emptible* Intelsat service. In particular, the use of very small rural stations in conjunction with an Intelsat space segment appears to be prohibitive. Moreover, the available capacity on relevant Intelsat satellites is not likely to be sufficient to meet the African mission requirement as defined at the beginning of this paper.

From the study of conventional ground systems it appears that the relation between such systems and an Afsat-type system is more complementary than competitive. In particular, it seems that, in most cases, for distances above 50–100 km for rural telephony and 100–180 km for inter-urban traffic, the satellite is more economical than terrestrial networks. Also, the satellite is advantageous for operational flexibility, deployment speed, alternative routing and vulnerability. Lastly, when a combined television–telephony usage is desirable, the satellite is more advantageous in all respects than ground-based facilities.

SPACE SEGMENT FINANCING AND ORGANIZATION

Basically it is considered that the space segment of the Afsat system will be financed under the usual aid schemes, i.e. by means of grants from the European Development Fund (E.D.F.) and from national Aid Agencies. In addition to such grants, privileged loans can be obtained from the European Investment Bank and similar national financial institutes (Caisse Centrale de Coopération, Kreditanstalt, etc.). The features of a privileged loan, it is recalled, are a low interest rate, a long duration and a long reimbursement moratorium. Obviously another substantial part of the funding will come from the investment of capital by the African telecommunication Administrations concerned.

However, other possibilities are also considered, such as investments or instalments by large users – governmental, African or otherwise – needing good telecommunications in Africa. An investigation of such users is currently being made by Eurospace under an E.S.A. contract.

This possibility, however, raises the problem of the legal link between such users and the participating African Administrations, i.e. the problem of the form of the African Organization capable of owning and operating the space segment. In the study, a two-tier scheme is proposed for this Organization, along the Intelsat precedent. This would comprise an Assembly of Parties to which African Administrations only would participate, and an Assembly of investors to which both African Administrations and investors and users would participate. Possibly investors and users could form a holding company that would represent them to the Organization. A Board of Directors would be responsible for administration. Initially it could subcontract the management tasks to a commercial company, as happened with Comsat in the early years of Intelsat.

CONCLUSION

The study that has been presented enables one to draw a number of conclusions.

(i) A dedicated African communication satellite system would bring a decisive contribution to the improvement of the continent's communications in terms of both increased capacity and better availability and quality of circuits.

(ii) The Afsat system is more economic than alternative space or terrestrial solutions and it is better adapted to the African environmental constraints.

(iii) The economic burden of Afsat appears to be compatible with African budgets.

(iv) External resources from aid and large users can bring major contributions to the funding and financing of the system.

A full scale feasibility study is still required, in particular to evaluate in detail the optimal use that each African country can make of the proposed Afsat system. This implies surveys in Africa to assess the existing networks, to collect information on tariff policies, to work out income assumptions, to propose combined conventional and space network solutions and establish their feasibility.

E.D.F. has already indicated its willingness to fund such a study provided that the application for funding is made by a credible African partner comprising English and French speaking countries, i.e. wider then U.A.P.T. To meet this requirement an Agreement of Intent for the continuation of the studies and the system's implementation will be open by the end of 1983 for signature by all the interested African countries. The group of signatories, it is felt, will form a motivated nucleus of responsible administrations capable of developing in due course into an operational African Organization. For these reasons Eurospace believes that all European countries and organizations concerned should give full support to this promising approach.

Clearly Europe cannot remain indifferent toward this project, which represents a considerable stake for its industry and is an exemplary cooperation exercise between high technology countries and a developing continent.

Discussion

S. METZGER (*Communications Satellite Corporation, Washington, D.C., U.S.A.*). Since the estimated space segment cost is $200 million while the Earth segment cost is $600 million to $800 million is it possible to use a higher powered, more expensive satellite to reduce the Earth segment cost, so that the overall cost is reduced?

What maintenance cost for Earth stations was assumed in Mr Demerliac's analysis?

What source of power was proposed for the remote Earth stations, and what was the estimated cost per station?

Y. DEMERLIAC. Yes, theoretically a more powerful satellite could reduce the Earth segment cost. However, to increase the e.i.r.p. (in 4 GHz/6 GHz) beyond the currently proposed value of 37 dBW would rapidly raise regulatory problems and, on the other hand, may affect the overall system optimization adversely.

The maintenance cost for Earth stations, disregarding operation and training costs, was assumed to be 3% of the installed equipment cost for class 1 and 2 stations and 6% for class 3.

Solar cell generators are proposed to power the remote Earth stations. The power cost per station varies widely according to the number of telecommunication channels and the local insolation. For countries with a high insolation the power cost for one s.c.p.c. channel is assessed to be $6000 plus $2500 per supplementary channel.

Phil. Trans. R. Soc. Lond. A **312**, 39 (1984)   [39]

Printed in Great Britain

Unisat: the British direct broadcast satellite system
[abstract only]

By P. L. V. Hickman

Space Division, British Aerospace p.l.c., Argyle Way, Stevenage,
Hertfordshire SG1 2AD, U.K.

Since the formation of the International Satellite Organization, communications satellites have been a clear, and until the present perhaps, the only, example of a commercial application of space. The money to finance these operations have come, for the most part, from state-owned institutions (the P.T.Ts) in the member states.

From the outset the U.K. Government made it clear that direct broadcasting by satellite in the United Kingdom would have to be financed by the private sector. This paper will briefly describe the steps that have so far been taken along this path by the providers of the satellite service and the problems that can be encountered.

Note. The abstract only is given here because, unfortunately, Mr Hickman's paper did not become available after the meeting.

Phil. Trans. R. Soc. Lond. A **312**, 41–54 (1984)   [41]

Printed in Great Britain

Mobile communications via satellite in the 1990s

By O. Lundberg

International Maritime Satellite Organization, 40 Melton Street, London NW1 2EQ, U.K.

The world's only satellite organization providing mobile communications on a commercial basis is the International Maritime Satellite Organization (Inmarsat). This paper reviews the origins of the organization and the needs of the shipping and offshore industry that led to its formation. The current system and its operations are described. The success achieved so far by Inmarsat in providing the satellite capacity for telephone, telex, facsimile and data communications to the maritime community makes it apparent that the system could also be used to provide capacity to the aeronautical community. Also, studies are now being made on future configurations of the system in which it may be possible to integrate a polar-orbiting satellite system, such as Sarsat–Cospas. Inmarsat is proceeding with the procurement of a new series of satellites that would come into operation from 1988. This paper reviews the enhanced capabilities that these new satellites will provide in the context of the requirements for mobile communications via satellite in the next decade.

Introduction

In the 1980s, satellites have become a mature technology. In the 1990s, the challenge, in part, may be the industrialization of space. More particularly, the challenge in the next decade will be to use that technology wisely, efficiently and in a way that serves the real needs of the end users in a vibrant marketplace.

Inmarsat sees its challenge in providing satellite **communications** capacity that best serves the mobile user in the international environment. This challenge is particularly important in view of the fact that Inmarsat is the world's only organization providing commercial mobile communications via satellite.

Although Inmarsat is a relatively young organization (it began operations in February 1982), the idea that impelled its creation is not. Indeed, maritime nations began to consider the possibility of a maritime communications satellite system as long ago as the launch of Early Bird in 1965.

The focus, then, of this paper will not be so much on the industrialization of space, but more on the future of mobile communications via satellite. Before embarking on that course, it would be appropriate to make reference to the origins of the maritime satellites.

Why Inmarsat was needed

Since the 1965 launch of the first commercial communications satellite, space-age technology has had a dramatic impact on the way the nations of the world communicate. Television, radio and telecommunications delivered via satellite now put people from around the globe in touch with each other in seconds. Until recently, however, this technological marvel was largely confined to the enjoyment of people on land, even though there has long been a need to extend the power of satellites to people at sea. It was for this reason that, soon after the launch of

Early Bird, maritime nations began to consider how satellites could be used to provide reliable communications to and from ships. They recognized that conventional radiocommunications on the m.f., h.f. and v.h.f. bands, although much improved over the years, cannot effectively meet all modern shipping requirements. They are still subject to the vagaries of ionospheric and other disturbances, as well as a lack of an adequate number of channels, making contact often difficult and sometimes impossible. Delays of many hours and even days occur, particularly in long-range communications. At times of maritime disaster, such delays have led in the past, and may in the future, to loss of life and property.

TABLE 1. INMARSAT MEMBER COUNTRIES AND SIGNATORIES
(DECEMBER 1983)

U.S.A.	Communications Satellite Corporation (Comsat)
U.S.S.R.	Morsviazsputnik (includes initial investment share of Byelorussian and Ukranian S.S.Rs)
U.K.	British Telecom
Norway	Norwegian Telecommunications Administration
Japan	Kokusai Denshin Denwa Co., Ltd
Italy	Ministero delle Poste e Telecomunicazioni
France	Direction Générale des Télécommunications
F.R.G.	Bundesministerium für das Post und Fernmeldewesen
Greece	Hellenic Telecommunications Organization (O.T.E.)
Netherlands	Netherlands P.T.T. Administration
Canada	Teleglobe Canada
Kuwait	Ministry of Communications
Spain	Compañía Teléfonica Nacional de España
Sweden	Swedish Telecommunications Administration
Australia	Overseas Telecommunications Commission (O.T.C.)
Brazil	Empresa Brasileira de Telecomunicacoes S.A. (Embratel)
Denmark	Post and Telegraph Administration
India	Overseas Communications Service
Poland	Office of Maritime Economy
Saudi Arabia	Ministry of Posts, Telegraphs and Telephones
Singapore	Telecommunication Authority of Singapore
China (P.R.C.)	Beijing Marine Communications and Navigation Company
Argentina	Empresa Nacional de Telecomunicaciones (Entel)
Belgium	Régie des Télégraphes et des Téléphones
Finland	General Directorate of Posts and Telecommunications of Finland
New Zealand	Post Office Headquarters
Bulgaria	Shipping Corporation
Portugal	Companhia Portuguesa Radio Marconi
Algeria	Ministère des Postes et Télécommunications
Chile	Empresa Nacional de Telecomunicaciones S.A. (Entel-Chile)
Egypt	National Telecommunications Organization (Arento)
Iraq	Republic of Iraq
Liberia	Republic of Liberia
Oman	Sultanate of Oman
Philippines	Philippine Communications Satellite Corporation (Philcomsat)
Sri Lanka	Overseas Telecommunication Service
U.A.E.	Ministry of Communications
Tunisia	Republic of Tunisia

In 1971, the World Administrative Radio Conference for Space Telecommunications allocated frequency bands to the maritime mobile satellite service. Four years later, the International Maritime Organization (I.M.O.) convened the first of three sessions of the International Conference on the Establishment of an International Maritime Satellite Organization. I.M.O. identified several reasons why such a system should be established: to relieve congestion in the m.f. and h.f. bands; to improve reliability, quality and speed of communications; to improve geographical coverage and continuous availability of services; to permit automation of radio-telephone and teleprinter; to cater for services not possible in the m.f. and h.f. bands, such as high speed data transmission; to improve distress, urgency and safety communications.

I.M.O. held the final session on 3 September 1976, at which time the Conference unanimously adopted the Convention and Operating Agreement on the International Maritime Satellite Organization. With its headquarters in London, Inmarsat came into existence on 16 July 1979 and began operations on 1 February 1982. Forty countries have now joined the international organization (table 1).

Inmarsat's purpose

According to its Convention, Inmarsat's purpose is 'to make provision for the space segment necessary for improving maritime communications, thereby assisting in improving distress and safety of life at sea communications, efficiency and management of ships, maritime public correspondence services and radiodetermination capabilities.' The Convention also says that Inmarsat shall act exclusively for peaceful purposes, that it is open for membership by all States and that ships of all nations may use the space segment.

Structure and financing

Established to 'operate on a sound economic and financial basis, having regard to accepted commercial principles', Inmarsat is financed by the Signatories to its Operating Agreement. The Signatories have been designated as such by the member countries. For the most part, the Signatories are national telecommunications carriers.

The Signatories come from the rich and the developing nations, from East and West. Those Signatories with the largest investment shares are from the U.S.A., Soviet Union, the U.K., Norway and Japan. The investment share is intended to reflect the actual use of the Inmarsat system. The investment covers the costs of the space segment and the operating and administrative costs of the organization. The total of all capital invested by the Signatories, plus compounded compensation at 14 % a year, is to be returned to Signatories as the system becomes profitable. Inmarsat could break even as early as 1984, much earlier than its original forecasts, in part because the number of users of the system is growing more rapidly than was expected.

Inmarsat has a three-tier organizational structure.

(1) The *Assembly* of Parties (or States) meets once every two years to review the activities and objectives of Inmarsat and to make recommendations to the Council. All member States are represented and have one vote each.

(2) The *Council* of Signatories is like a corporate board of directors; it is Inmarsat's policy-making body. It consists of 22 Signatories: 18 with the largest investment shares and four others elected by the Assembly on the principle of a just geographical representation and with due regard for the interest of developing countries. The Council meets at least three times a year and each Signatory has a voting power equal to its investment share.

(3) The *Directorate* achieves the day-to-day activities of the organization. The Director General is Inmarsat's Chief Executive Officer.

The Inmarsat system

The Inmarsat satellite system is somewhat different from those of other satellite organizations which provide a fixed service. Inmarsat interfaces directly to the end user's telephone on the mobile side of the satellite. It extends the terrestrial telecommunications networks to the oceans of the world. When someone on board a ship picks up the telephone, he is, in effect, using the ultimate in cordless telephones.

The maritime satellite system has three major components: the satellite capacity leased by the organization, the coast Earth stations and the ship Earth stations (figure 1).

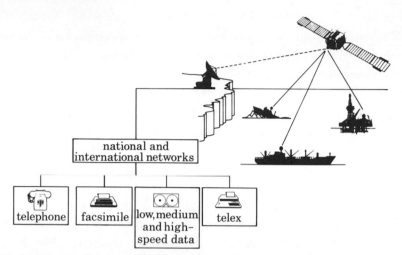

FIGURE 1. The Inmarsat system comprises three main elements: the satellites, the coast Earth stations and the ship Earth stations. The coast Earth station provides the link with the national and international telecommunications networks. The broken line represents transmissions to the satellite at 6 GHz and reception from the satellite at 4 GHz; the solid lines represent transmissions at 1.6 GHz and reception at 1.5 GHz.

The nerve centre of the system is the operations control centre (o.c.c.) at Inmarsat's headquarters. The o.c.c. is connected directly by its own ship Earth stations to the Atlantic and Indian Ocean satellites, and so to all coast Earth stations around the world. Operating continuously, it coordinates a wide range of activities. Should a serious problem arise with one satellite, putting it out of commission, the o.c.c. would be responsible for taking the necessary steps to transfer traffic to a spare satellite in orbit. The o.c.c. also handles all commissioning applications from ships that have just installed Inmarsat terminals.

The satellites

The Inmarsat satellites are in geostationary orbit, 36 000 km above the equator, over the Atlantic, Indian and Pacific Oceans, and provide near-global coverage (figure 2). There are both operational and back-up satellites. Thus, if one satellite should fail, another spare in orbit will be able to take over immediately.

From December 1983, Inmarsat had leased two Marisat satellites, each with a capacity of about 10 telephone channels, from the Marisat Joint Venture (the principal shareholder of which is Comsat General of the U.S.A.); a Marecs A satellite, with a capacity of about 40 telephone channels, from the European Space Agency; and maritime communications subsystems (m.c.s.), each with a capacity of about 30 telephone channels, on the Intelsat V F-5 and F-6 satellites. In the next year or so, it plans to lease capacity on other satellites. The satellite configuration is given by ocean region in table 2.

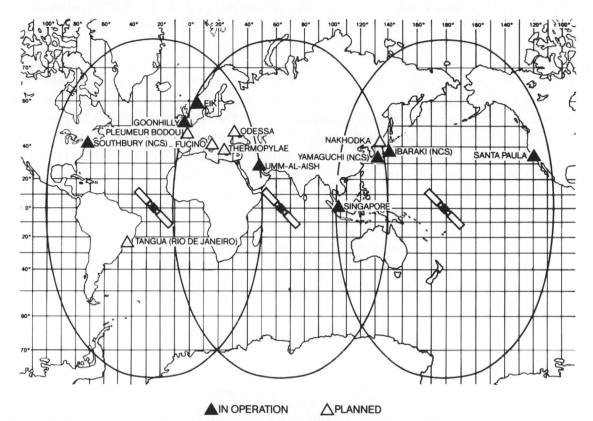

▲ IN OPERATION   △ PLANNED

FIGURE 2. The Inmarsat satellites provide near global coverage (to about the 75° latitudes) so that a ship can communicate via satellite virtually anywhere. As of 1 December 1983, there were eight coast Earth stations in operation with several more scheduled to joint them in 1984. Three of these are Network Coordination Stations (N.C.S.), which assign telephone channels on demand to requesting ship Earth stations.

TABLE 2. SATELLITE CONFIGURATION

ocean	satellite	status
Atlantic	Marecs A	operational
	Intelsat V MCS-B	in-orbit spare
Indian	Intelsat V MCS-A	operational
	Marisat F-2	in-orbit spare
	Intelsat V MCS-C	available late in 1983
Pacific	Marisat F-3	operational
	Marecs B2	available early in 1985

Coast Earth stations

The coast Earth stations provide the link between the satellites and the telecommunications networks ashore. The coast Earth stations are owned and operated by Signatories, who are also responsible for the landline connections to the public switched telephone and telex networks. A typical coast Earth station consists of a parabolic antenna about 11–14 m in diameter, which is used for transmission of signals to the satellite at 6 GHz and for reception from the satellite at 4 GHz. The same antenna or another dedicated antenna is used for L-band transmission (1.6 GHz) and reception (1.5 GHz) of network control signals. Each coast Earth station provides, as a minimum, telex and telephone services. As well, three coast Earth stations – at Southbury, Connecticut, and at Yamaguchi and Ibaraki in Japan – serve as network co-ordination stations, which assign telephone channels, on demand, to ship Earth stations.

As of October 1983, eight Inmarsat coast Earth stations were in operation around the world, and their number is expected to more than double by the end of 1984.

The additional coast Earth stations will shorten the terrestrial distance a call has to travel to and from ships and offer the opportunity to reduce user charges. The cost of placing a call varies from country to country, as one might expect, since end-user charges for Inmarsat services are set by telecommunications administrations. For ship-to-shore calls, the Signatory operating the coast Earth station bills the user's designated accounting authority. Billing for shore-to-ship calls is the responsibility of the telecommunications administration providing service via Inmarsat, which in turn earns its revenues by billing the Signatories who operate the coast Earth stations for the use made of its satellite system.

Ship Earth stations

The ship Earth station, which puts the user on ship in instant contact with the rest of the world, consists of two parts, the hardware above deck and that below. That above consists of a parabolic antenna, typically between 0.8 m and 1.2 m in diameter, housed in a fibreglass radome, which protects the dish from the harsh maritime environment (figure 3). The antenna is mounted on a stabilized platform, which enables the antenna beam to remain pointed at the satellite regardless of ship course or movement. Signals are transmitted to the satellite at 1.6 GHz and received at 1.5 GHz.

FIGURE 3. The ship Earth station, which enables a ship to communicate with other ships or with shore-based subscribers, consists of equipment both above and below deck. Above deck is a parabolic antenna on a stabilized platform covered in a radome. The radome on the bulk carrier *Ambia Fair* can be seen here as a white, distinctive mushroom shape mounted on a pedestal above the bridge. (Photograph by Skyfotos.)

The equipment below the deck consists of telex and telephone and a variety of optional equipment for facsimile, data and slow-scan television. Low and medium-speed data transmission are available via a voice channel at up to 4.8 kbit s^{-1} in both directions. With a specially prepared satellite channel in the ship-to-shore direction, high-speed data transmission at up to 56 kbit s^{-1} is available. Inmarsat has also agreed to make satellite capacity available in the ship-to-shore direction for a very high speed data service (v.h.s.d.), which would allow information to be transmitted at rates of 1 Mbit s^{-1} or more.

Calls are placed to and from ships in virtually the same way – and just as quickly – as one would make a call from the home or office. When a user on board a ship places a telephone call, he pushes a request button on his communications equipment below deck, which sends a signal via satellite to a network coordination station (n.c.s.). The n.c.s. then assigns a voice channel from a 'pool' of available circuits for use by the ship for that call. Telex channels, however, are assigned by the coast Earth station. In both cases, channels are assigned automatically and in a matter of seconds. The user then dials the number he wants. Each ocean region has the equivalent of a country code.

Inmarsat gives distress signals priority access to the spare segment. For distress calls, a special button on the ship station may be pressed to provide immediate telex or telephone communication to the desired coast Earth station, which then puts the ship into contact with the appropriate rescue coordination centre.

Ship Earth stations can be purchased from 10 different manufacturers around the world. Prices are competitive at about U.S. $35 000, which is about half as much as one cost two or three years ago. This technology will continue to evolve, as manufacturers of ship Earth stations are producing a new generation of equipment that is smaller and easier to use than earlier models. Until now, the antennas above deck have been of parabolic design, but now there is one manufacturer seeking approval from Inmarsat for a phased array antenna. Most of the current systems below deck are modular in design and can easily be upgraded by the addition of new capabilities. A typical unit below deck consists of a microcomputer with a visual display unit (v.d.u.) and alphanumeric keyboard, hard copy printer and telephone with a modem.

For the time being and probably until the early 1990s, the standard A ship Earth station, an analog terminal, is likely to remain the all-purpose workhorse of the Inmarsat system. Development work is, however, proceeding on other possible standards to meet the needs of high- and low-volume-communication users. One of these could be a multi-channel variant of the standard A, which currently has a one telephone–one telex channel capability. Inmarsat recently approved new technical requirements that would permit up to three simultaneous telephone or telex calls, a capability that would be of interest to heavy users of the system such as the oil industry. Other standards would include a smaller, digital version of Standard A and an even smaller terminal, which would be used for telex or low-speed data communications. These new standards could reach the market at about the same time as Inmarsat's second-generation satellite system, in 1988.

APPLICATIONS AND USERS

The number of ships, offshore drilling rigs and others using the satellite system increased by about 60 % in Inmarsat's first year of operations (figure 4). On 1 December 1983, there were about 2200 users, including oil tankers, liquid natural gas carriers, offshore drilling rigs, seismic survey vessels, fishing boats, cargo and container vessels, passenger ships, ice breakers, tugs, cable-laying ships and even a replica of a Viking ship. The maritime satellite system is also used by research teams in the Antarctic, a weather station in northern Greenland and, on an exceptional basis, by oil production platforms. As might be expected from such a range of users, the range of applications of satellite communications is equally wide. Some examples are given.

(1) Offshore oil rigs equipped with satellite terminals can transmit well log data to land-based computers for rapid analysis.

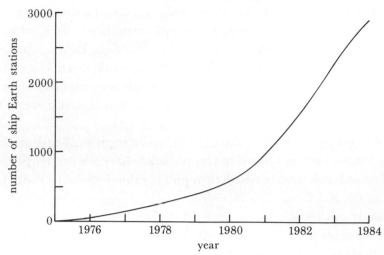

FIGURE 4. In its first year of operations, Inmarsat witnessed a 60% increase in the number of ships fitted with ship Earth stations, despite the prolonged recession in the shipping industry.

(2) Ship operators can use the system for the rapid transmission of data on engine perform-ance, fuel consumption, position-reporting, weather conditions and provisioning. Their vessels can obtain berthing and scheduling information in advance of arrival in port. If there are delays in port, the ship can adjust its estimated time of arrival or be diverted to another port.

(3) If a vessel breaks down at sea, engineering drawings can be transmitted by facsimile to enable repairs to be made. If necessary, spare parts can be ordered by satellite and be waiting in the next port to minimize the time spent waiting for the repairs to be done.

(4) Ships of a particular fleet or national group can receive a common telex message through Inmarsat's facility for shore-to-ship group ('broadcast') calls. Facilities also exist for group calls to ships in a defined geographical area within each ocean region, which is particularly useful for broadcasting weather forecasts and navigational warnings.

THE FUTURE OF MOBILE COMMUNICATIONS VIA SATELLITE

Since 1979, Inmarsat's preoccupation has been to establish its satellite system. Now, with a proven, reliable system in place, one that is attracting a rapidly growing number of users, Inmarsat has been devoting an increasing amount of its time to the future. In the maritime world, a milestone will be reached in 1990 when the Future Global Maritime Distress and Safety System (F.G.M.D.S.S.) is introduced by the International Maritime Organization (I.M.O.).

Future Global Maritime Distress and Safety System

F.G.M.D.S.S. will also be a milestone for Inmarsat, because one of its major objectives is to provide improved facilities for distress calls and communications to improve safety of life at sea. The immediacy and reliability of Inmarsat communications have already led some countries to include the ship Earth station as an alternative main transmitter for mandatory carriage on ships covered by the I.M.O. Safety of Life at Sea (SOLAS) Convention. Many more countries are expected to follow suit as I.M.O. prepares its new SOLAS requirements and its proposals for the F.G.M.D.S.S. According to I.M.O., the F.G.M.D.S.S. is intended to be 'a comprehensive system to improve distress and safety communications and procedures which, in conjunction with a coordinated search and rescue infrastructure, will incorporate recent technical developments'.

One objective of the future system is to enable any properly equipped ship to achieve automatic distress alerting and be located with minimum delay. Although most countries have some sort of search and rescue facilities for aiding ships at sea, these facilities vary and their efforts are not coordinated on a global basis. Hence, the system would also provide for the establishment of rescue coordination centres where they do not exist and for a set of procedures for responding to distress alerts and for making rescue missions. As well, it is expected that most rescue centres will be equipped with satellite communications so that they can have direct access to the Inmarsat system and will, as a consequence, be able to contact vessels similarly equipped in seconds in the event of a distress or emergency at sea. In November 1983, Argentina became the first country to equip a rescue centre with maritime satellite communications.

Inmarsat is expected to play a major role in the F.G.M.D.S.S. I.M.O. has said 'The F.G.M.D.S.S. will use both satellite and terrestrial communications. Satellite communications will be provided by Inmarsat. A distress capability for alerting by satellite EPIRB will be provided by Inmarsat geostationary satellites as well as polar orbiting satellites. Terrestrial communications will use frequencies in the m.f., h.f. and v.h.f. bands. Terrestrial communications will no longer use Morse code radiotelegraphy but will employ digital selective calling, radiotelephony and narrow band direct printing.' All equipment carried on ships will be designed for simple operation and will be largely automated.

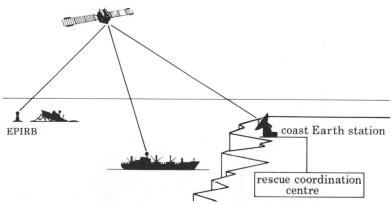

FIGURE 5. An emergency position-indicating radio beacon is a transmitter fitted on a buoy designed to float free from a sinking ship and to transmit a distress signal via satellite to a coast Earth station, which would then relay the message to the nearest rescue coordination centre. The Inmarsat system has been used in recent trials of such EPIRBs.

The satellite EPIRB (emergency position-indicating radio beacon) will be capable of sending a message via satellite by giving the ship's identity number and its position coordinates (figure 5). It could be carried on a ship or lifeboat, fitted in a buoy designed to float free and be activated should a ship sink. I.M.O. is preparing the operational requirements, while the International Radio Consultative Committee (C.C.I.R.) is developing the technical provisions for a satellite EPIRB system. In 1982–83, a working group of the C.C.I.R. coordinated a series of trials using the Inmarsat system aimed at developing the technical specifications for a EPIRB, operating at L-band, the same frequency bandwidth used by Inmarsat.

Polar-orbiting satellites

There is, of course, another search and rescue satellite system operating, called Sarsat–Cospas, which has been developed by Canada, the United States, France and the Soviet Union. The Soviet Union was the first to launch a satellite under this scheme, the Cosmos 1383, in June 1982. Two months after its launch, the satellite picked up signals transmitted from an aircraft that had crashed in northern British Columbia. By measuring the Doppler shift in the received signals, a rescue team was able to pinpoint the location of the aircraft. The speedy rescue that followed resulted in three lives being saved. Since then, two Cosmos satellites and an American NOAA satellite carrying similar payloads have been credited with saving more than 60 lives.

These satellites operate in a polar orbit up to 1000 km above the Earth and detect signals in the 121.5 and 243 MHz band, the international distress frequencies, as well as in the 406 MHz band, which was recently allocated specifically for low power emergency beacons capable of being detected and located by satellites.

There are pros and cons to be said for satellite EPIRBs operating at 406 MHz or L-band. For its part, Inmarsat has said that it does not intend to carry a 406 MHz payload on its second-generation satellite system, as we have not yet identified any customer who is willing to pay for the extra expense involved. In any event, the trials using our satellites have shown that the emergency beacons can operate effectively at L-band.

I.M.O. has said that it would like to see a single international distress frequency for satellite emergency beacons. That would certainly be desirable from the perspective of the user whose equipment is likely to be cheaper than would be the case if it had to operate at two or more frequencies. It may well be that the optimum system would integrate both polar-orbiting and geostationary satellites in one scheme (figure 6). Polar-orbiting satellites could carry additional payloads as well, perhaps capabilities for navigation and thin-route communications, which would be especially valuable over the polar regions.

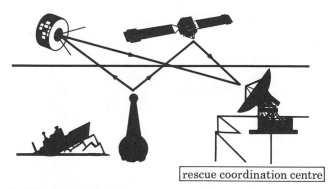

rescue coordination centre

FIGURE 6. In the next decade, it is possible that a polar-orbiting satellite system such as Sarsat–Cospas could be integrated with the geostationary Inmarsat satellite system. This combination would principally be used for distress alerting, but polar-orbiting satellites could carry other payloads as well, including thin-route communications for use over the polar regions.

It was recently decided by the Sarsat–Cospas member countries, which now also include the U.K. and Norway, to carry on with this demonstration satellite programme until the end of the 1980s. Until then, member countries will be giving some thought as to what sort of institutional structure should be established for managing this programme on a continuous basis. It has been suggested that one option would be to have Inmarsat procure and manage such a system. Probably among the reasons that Inmarsat might be considered are that it has the right sort of institutional structure, that it has the mandate to provide for distress and safety services and that Inmarsat itself has been participating in trials, organized by I.M.O. and the C.C.I.R., of satellite emergency beacons. Some of the Sarsat–Cospas nations have also been involved in these trials.

Inmarsat may be willing to manage a search and rescue satellite programme, if it could be financed. Someone would have to pay for such a programme; at present, it is the taxpayers in the participating countries. If the programme were managed by Inmarsat, it would be possible to spread the burden to all countries that benefit. While it's true that Inmarsat was created to provide such services, a formula for funding them is still needed.

Aeronautical communications

Although Inmarsat was created to serve maritime requirements, other possible applications were foreseen at the international conference that conceived the organization. One of these applications, the feasibility of which Inmarsat is exploring, relates to aeronautical communications via satellite. The international conference recommended that 'arrangements should be made to undertake at an early date the study, without prejudice to programmes in planning, of the institutional, financial, technical and operating consequences of the use by Inmarsat of multi-purpose satellites providing both a maritime mobile and an aeronautical mobile capability'.

The notion of aircraft using satellite communications is not a new one. In 1968, I.C.A.O. established a panel of experts to deal with the Application of Space Techniques Relating to Aviation (ASTRA). Over the next four years, the panel studied the technical characteristics for an aeronautical satellite system. Following a recommendation of an Air Navigation Conference held by I.C.A.O. in 1972, representatives from Canada, the United States and the European Space Agency set up the Aeronautical satellite (Aerosat) programme. Aerosat was intended to follow an international programme for research and development of an aeronautical satellite. The experimental Aerosat satellite was to determine the desired characteristics of an operational aeronautical satellite for mobile communications and position determination. The satellite was to be launched in 1979–80, but it never reached the launch pad. By the end of the decade, the aviation industry was in the financial doldrums, and airlines were reluctant to spend money on an expensive, dedicated satellite system; funding for the programme ceased. Fortunately, the interest in aeronautical satellite communications did not diminish. In fact, I.C.A.O. and the International Air Transport Association (I.A.T.A.) had even participated in some of the early discussions that eventually led to the establishment of Inmarsat.

In late 1982, I.C.A.O. said it intended to draft technical specifications for a future satellite system that could serve the mobile communication needs of international civil aviation. The service as envisaged by I.C.A.O. would be designed to improve air–ground communications and would make use of low-speed digital data transmission on the L-band frequencies (specifically, the aeronautical R-band) already exclusively allocated for the aeronautical mobile service by the I.T.U.

In the months that followed, Inmarsat continued its intensive discussions with I.C.A.O. and I.A.T.A. and others, which were aimed at defining the aviation community's requirements and exploring institutional arrangements wherein the Inmarsat system could be shared for aeronautical applications. Prompted by I.C.A.Os expression of interest in a satellite communica-

tions facility, Inmarsat's 14th Council session in May 1983 approved an amendment to a request
for proposals, which was released in August 1983, for the second-generation Inmarsat satellite
system. The amendment meant that Inmarsat could provide an aeronautical capability with
its new series of satellites, the first of which is scheduled for operation in late 1988. The amendment principally involved an extension of the upper end of the L-band bandwidth by about
1 MHz, and some changes to the C-band feeder link frequency bandwidths. The fact that
Inmarsat's second generation satellites will incorporate a small part of the aeronautical L-band
does not mean that Inmarsat is committed to providing an aeronautical service, nor that the
aeronautical community is committed to using it. At this stage, it merely means that Inmarsat
could provide one, subject to detailed consultations with the aeronautical community.

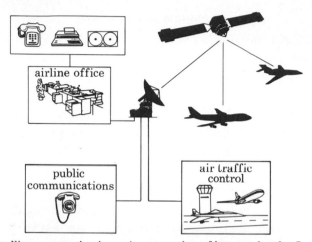

FIGURE 7. Aeronautical satellite communications. An expression of interest by the International Civil Aviation
Organization in aeronautical communications via satellite prompted Inmarsat to amend its request for
proposals for a new series of satellites to come into operation in 1988. The amendment means that aircraft
could use the same satellite system now used by the maritime community.

The adjacency of the maritime and aeronautical frequency bands – and thereby the possibility of sharing the same satellites – would seem to be a powerful argument for cooperation
between Inmarsat and the aeronautical community. It means that Aerosat capabilities could
be provided at a cost orders of magnitude less than earlier assumed. While cooperation between
the two user groups would seem to be logical, all solutions must be driven by user requirements.
One requirement is for improvement in present radio links. Very high frequency radio leaves
large parts of the globe uncovered and h.f. suffers from well known propagation peculiarities
that make the medium unsuitable for modern communications. Additionally, existing links
sometimes suffer from congestion and crowding of the air waves. Satellites, on the other hand,
offer high quality, reliable and virtually instantaneous air–ground communications. Aircraft
could use satellite communications for air–ground data links for air traffic control – particularly
across the north Atlantic, where airline traffic is the heaviest – and for transmission of flight
data and weather reports (figure 7). Satellites also have a communications range that extends
to almost one-third of the world. Three would provide near global coverage, a fact that has
considerable implications for system design; a satellite system would probably be far less
expensive than any new terrestrial solutions.

Although a satellite communications terminal for civil aircraft has yet to be developed,
Inmarsat is making various studies that could facilitate its development. One of these studies

is being made under a contract that is aimed at defining certain technical aspects of providing aeronautical communications via the Inmarsat system. The contract encompasses the following areas: propagation effects on aeronautical satellite links; antenna systems; modulation and coding methods; overall aircraft Earth station (avionics) integration.

Another study is expected to lead to development of a compact, lightweight and relatively inexpensive type of Earth station, which could be used by ships or aircraft. This terminal, known as standard C, is initially intended to provide maritime users with medium speed digital telex-type communications. Standard C will be characterized by a low-gain, quasi-omni-directional antenna. Preliminary estimates indicate that highly reliable communications could be achieved with a G/T in the range of -17 to -26 dB K^{-1} and with a mobile terminal e.i.r.p. of about 20 dB (with respect to 1 W). An overall terminal mass of 15 kg or less is envisaged. Inmarsat estimates that standard C terminals could cost less than U.S. \$10 000, although an aircraft terminal of similar capabilities would require integration with other aircraft avionics and may prove to be more expensive as a result. Although a voice capability for such a satellite terminal is likely to be further into the future, Inmarsat is studying possible digital voice coding methods compatible with low data rates.

With development of a suitable terminal, aircraft could start using Inmarsat's existing satellite system on an experimental basis. It is more likely, however, that the aeronautical terminals would not start appearing in any substantial numbers until after 1988, when Inmarsat's second generation satellite system comes into operation.

SECOND-GENERATION SYSTEM

The satellites composing Inmarsat's first-generation satellite system will near the end of their design lives in the late 1980s, by which time the available capacity in the Atlantic Ocean region is expected to become saturated. In August 1983, Inmarsat issued a call for tenders on a new series of satellites, which could cost several hundred million dollars. Inmarsat requires new satellites from 1988 to handle expected increases in the numbers of users of its system and to offer new services.

Inmarsat's request for proposals, which was sent to leading satellite suppliers around the world, says that bidders could offer spacecraft to be leased or purchased by Inmarsat. Depending on which option is selected, Inmarsat could be ordering as many as nine satellites, which would be launched into geostationary orbit over a three-year period starting in the second quarter of 1988. The spacecraft must be compatible with more than one launcher of six listed in the request for proposals. Of the six, one (Ariane) is European, one (Proton) is Soviet and four (the Space Shuttle, Thor Delta, Atlas Centaur and Titan) are American.

Inmarsat's second-generation system is expected to have many more times the capacity of its existing system. The request for proposals calls for spacecraft with a minimum of 125 telephone channels, compared to the 40 channels provided by the Marecs satellite, the one with the greatest capacity in the current system. The channel capacity could be expanded to about 250 channels by using various techniques, notably carrier suppression. In the ship-to-shore direction, the spacecraft will provide four separate channel paths, which will allow better use of available capacity for working with new smaller types of ship Earth stations under development for the late 1980s, as well as with EPIRBs and aircraft.

Bidders are required to send their proposals to Inmarsat no later than 2 April 1984. The successful bidder or bidders will be chosen by the first quarter of 1985 and then, under the purchase option, will have 36 months to deliver the first in the new series of satellites. If it decides to purchase, rather than lease the satellites, Inmarsat would place contracts for the launch of the satellites.

As noted earlier in this paper, the request for proposals provides for the possibility that the second-generation system may be used for satellite communications to and from aircraft, as well as for new, smaller and more advanced ship Earth stations to be introduced in the future. The number of users of the system is expected to exceed 10 000 by the mid-1990s, and it may be even more than that if Inmarsat is used to serve the aeronautical community and to manage a search and rescue satellite system.

COOPERATION OR FAILURE

The realization of such a multi-purpose system in the 1990s will depend on the support and participation of all nations concerned. Any international system such as that required to support international aeronautical or maritime transport must be open for use by all nations on a non-discriminatory basis. This is not really a political requirement but is primarily based on common sense and a recognition of the fundamental role such a system would play in supporting international transport and improving safety.

The broad international base of Inmarsat suggests that it should be possible to develop institutional arrangements reasonably quickly for search and rescue satellites and for aeronautical satellite communications, following the model already set for maritime satellite communications. Indeed, steps are being taken in this direction already, following a decision by Inmarsat member states during the third Inmarsat Assembly in October 1983. The Assembly requested the Director General to study what amendments would be required to the Inmarsat Convention and Operating Agreement, to put the provision of aeronautical services via Inmarsat on a sound institutional footing. In addition, for an international system like Inmarsat to be successful, there is a need for technical standards to be agreed to achieve economies of scale and interconnectivity. Mobile users need only one system, one piece of equipment that can be used all over the world.

The next decade should be an exciting one for international mobile satellite communications. It is possible that the next decade will see such communications provided by a multi-purpose satellite system, in which geostationary and polar-orbiting satellites are integrated. Such a system, different in many respects from that of the 1980s, will probably be the only way in which various mobile communities can be served in an economic fashion.

Phil. Trans. R. Soc. Lond. A **312**, 55–66 (1984)   [55]
Printed in Great Britain

Earth stations for fixed and mobile services

By J. V. Charyk and S. Metzger

Communications Satellite Corporation, 950 *L'enfant Plaza, SW,*
Washington, D.C., 20024, *U.S.A.*

This paper discusses the evolution of fixed and mobile stations during the past two decades and extrapolates trends into the future.

It is shown that for the various communications satellite systems discussed, i.e. international, domestic, maritime and direct broadcast, the terminal designs are controlled not only by technical factors but also by economic, organizational and political factors.

The net effect of these combined forces, while yielding less than optimum designs from a technical or economic viewpoint, has made for a remarkable growth in twenty years. From an experimental–operational system of four stations in 1965, which used a satellite with 10 W of effective radiated power, to, by 1990, millions of home television stations using a satellite of 200 kW effective radiated power per channel.

1. General introduction

Commercial satellites communication service began 18 years ago with the launch of Early Bird (Intelsat I). This paper will describe the evolution of the Earth stations used in the Intelsat system, and will focus on the technical, political and operational factors that have influenced their design. The outstanding success of the Intelsat system led to the use of communication satellites for domestic, maritime and direct broadcast use as well. The growth of these systems has exceeded all expectations. From four stations in 1965, there are now 650 stations working with the Intelsat satellites. These stations are either using Intelsat capacity directly for international service or are part of domestic and regional systems that have leased capacity from Intelsat. Domestic systems in the United States of America total over 500 transmit–receive Earth stations, with antennas measuring from 1.2 m to 30 m diameter and traffic per station ranging from one circuit to thousands. There are approximately 10 000 stations with four to five metre antennas for cable television receive-only head ends, and about 400 000 to 500 000 receive-only antennas for 2.5–5 m diameter used by individuals for television reception in the 4 GHz band. The Intersputnik, Canadian, and Indonesian satellite systems have been operating for a number of years with a total of hundreds of Earth stations. Recently, India initiated its own system, and by the second half of this decade, ten other countries are expected to launch national systems.

More than 2000 commercial ships communicate via satellite; by the end of the 1980s, there should be millions of home receivers for direct broadcast satellite television reception.

The major thrust in the design of large fixed stations has been to increase the telephone circuit capacity per station, at a decreased cost per circuit. In addition, new services have been and are being developed to explore the special capabilities of satellite transmission, in particular the development of maritime and aeronautical services, direct broadcast television services, and new business services permitting wideband transmission with small Earth stations.

2. FIXED INTERNATIONAL SERVICE STATIONS

(a) Overall considerations

When fixed international-service Earth stations were introduced in the 1960s, they differed from terrestrial radio relay stations mainly in degree rather than in kind. Because of the thousandfold increase in distance between satellite and Earth station, compared to a radio relay hop, the Earth station's antenna had a diameter ten times greater. The antenna facing the cold sky permitted the use of cryogenically cooled receivers with noise temperatures of tens of degrees rather than thousands. The transmitter output amplifier was in the kilowatt range rather than in the few watt range. Fortunately, the 40 dB fading allowance needed for terrestrial relays could be decreased to a few decibels. Also a transcontinental relay system must allow for the build-up of noise in 100 receivers, while a satellite system includes only two.

In all previous communications systems, the owners and operators of the system could balance the design of the various elements of the system to minimize the cost of the system as a whole. But in Intelsat, a unique political problem arose because each country's earth station was owned completely by that country, whereas each country shared the cost of the satellites based on that country's share of the total traffic. About 20% of the 109 member countries generate approximately 80% of the traffic. A country with 1% of the traffic today carries about 650 circuits, but 80% of the countries have fewer circuits, typically a few dozen. The largest users are the United States, with about 16000 circuits, and the United Kingdom with approximately 8500 circuits.

The major element of cost for the high traffic countries is the space segment, while for the lower user countries, the Earth station cost predominates. The lower traffic countries have no interest in increasing their Earth station costs by purchasing a new multiplex terminal that would use the power and bandwidth of the satellite more efficiently, thereby increasing the satellite telephone channel capacity and reducing the cost per channel. Their space segment cost might be only one tenth the cost of their Earth station, without the addition of the multiplex equipment. The large user would have the reverse problem and would be delighted to make the change. But since the large and small users communicate with one another, they must have the same type of multiplex. Initially, all countries used F.D.M./F.M., but by the early 1970s the differences in traffic were so great that single channel per carrier (s.c.p.c.) was introduced for use in light traffic streams. Beginning with Intelsat IV, a special form of s.c.p.c. was initiated for demand assignment (SPADE), which resulted in still higher efficiencies by making frequencies from a pool available, shared by all on demand.

During the 1980s, a third form of multiplex–modulation will be introduced, T.D.M.A., which is best suited for the moderately high traffic streams shared by a number of countries.

The Intelsat system charges the same price per telephone channel to a country using six circuits as to one that uses 6000 circuits. This situation is at odds with pricing policies in all other businesses. Further, since the heaviest traffic stream in the Intelsat system is across the North Atlantic, it is technically possible to serve this stream with a relatively small satellite of the Delta class, about half the size of the Atlas Centaur or Ariane class used to launch the Intelsat V. By taking advantage of the proximity of the major Earth stations in Western Europe, and of those in Eastern United States and Canada, 2.5° beams might be used. With an efficient multiplexing system, about 135000–180000 two-way circuits could be obtained, compared to the 35 000 circuits in an Intelsat VI, satellite, which weighs more than 3.5 times as much and costs that much more.

However, another satellite would be required to tie together all of the above countries with the others of the Atlantic region. With the heavy traffic skimmed off, the remaining traffic would cost more.

The policy for the Intelsat system has always been to emphasize its global responsibilities, especially to the lesser developed countries. This has resulted in the same Intelsat charge per telephone channel to all countries regardless of the number of channels leased. It has also resulted in the design of universal satellites rather than specialized satellites favouring the North Atlantic countries at the expense of those in South America and Africa.

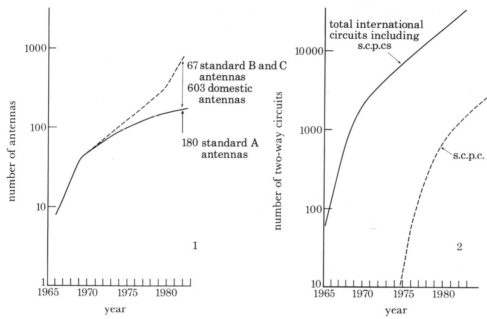

FIGURE 1. Growth in the number of Earth stations. The numbers shown are for the end of each year.

FIGURE 2. Traffic growth.

Similarly, the decision to adopt a geostationary system for the Intelsat system, rather than one at medium altitude, resulted in Earth stations requiring one antenna (and related equipment) instead of two. The decrease in cost by almost one half was of great significance to the smaller traffic countries to whom Earth station cost is predominant, being many times the space segment cost.

The extensive difference in the amount of traffic has resulted in changing from the initially universal standard A station (about 30 m diameter) to the introduction of standard B station (11 m diameter). Both of these stations are in the C-band, and use the frequencies from 5925 to 6425 MHz for transmission from Earth station to satellite, and from 3700 to 4200 MHz from satellite to Earth station. Standard C is a recent K-band station, with a 14 m diameter antenna (14–14.5 GHz from Earth station to satellite and 10.7–11.7 GHz for the reverse). In addition, two new series of stations for the new Intelsat Business Service have been approved. Standard E-1, E-2, and E-3 operate in the K-band with diameters of 3.5, 5 and 7 m, respectively; standard F-1, F-2, and F-3 operate in C-band with diameters of 5, 7, and 9 m, respectively. Thus the system is gradually evolving to match better the wide variety of new requirements that have developed with the advent of higher powered satellites and the use of new frequencies in bands free from terrestrial radio relay systems.

The change in the nature of Intelsat Earth stations over the years may be seen in figure 1, which shows that the initial users were the heavy traffic countries using the system exclusively for point-to-point communications with each other. In the early 1970s, lower volume traffic stations entered the system, and it was more economical for them to use the standard B station By the mid-1970s, transponders were leased for domestic use, for the most part with 5.5–11 m antennas. These now constitute the major part of Intelsat's 800 antennas, but the revenue from these leased transponders amounts to only about 8 % of Intelsat's revenue. Similarly, figure 2 shows the growth of s.c.p.c. circuits in use for international traffic only. It appears to follow the same curve as the growth of Intelsat's total traffic (F.D.M./F.D.M.A.) but delayed by about ten years. However, at present it constitutes only about 4 % of the total traffic.

(b) Antennas

In early 1964, before the formation of Intelsat, Comsat made a decision to launch an 'Early Bird' satellite, which later became known as Intelsat I after Intelsat was formed in mid-1964. This satellite was the heaviest (39 kg) that could be placed in a geostationary orbit with the Delta launch vehicle available at that time. The satellite could support only a single 6 W travelling wave tube, which, with the two carriers required, one for each direction of transmission, resulted in an effective radiated power of 10 W per carrier. To achieve a worthwhile traffic capacity with this low power, the receiving antenna had to be as large as possible. At that time, that meant a 25.9 m diameter antenna, which, if a 50 % efficiency and a 50 K noise temperature is assumed ($G/T = 40.7$ dB K^{-1}), provided 240 two-way telephone circuits. Comsat had purchased A.T.&Ts Andover, Maine, station, which used a horn antenna (see figure 3) with approximately this performance capability. A.T.&T. had chosen the horn design because it had used thousands of such antennas, but with a 2.4 m diameter, for its radio relay installations. The antenna had a simple geometry, with readily calculable parameters. A.T.&T. had built successfully a 6.1 m version for the Echo project. Therefore, it felt confident that a 20.7 m version could be built. It proceeded to build two antennas of this type, one for Andover and one for the French station at Pleumeur-Bodou, in 1962. The German station at Raisting had a 25 m Cassegrain antenna, and the United Kingdom built a 25.9 m front-fed antenna. Of these three designs, the Cassegrain has been proven to be the best choice, considering performance and price. It is now used almost universally.

Early antennas were copied from designs that did not require maintenance during operation. Intelsat's emphasis on the need for ready access to all equipment at any time stimulated changes in the mechanical design of the antennas from the kingpost (see figure 4, which shows the antenna at Brewster (1966)), which unfortunately required a technician to perform maintenance work on equipment at the base of the feed, doing so at the elevation angle of the antenna. For Comsat's antenna in Hawaii, this entailed working at a 45° angle, a most uncomfortable position. To overcome this problem, the following design incorporated an elevated room that remained horizontal regardless of the elevation angle (see figure 5, which shows the antenna at Etam (1968)). The wheel and track design permitted the antenna to be built on top of the building, which eliminated the concrete pedestal of the previous designs and thereby reduced costs (see figure 6, which shows an antenna in Alaska (1970)). Finally, the thrust to minimize the number of operating personnel and to reduce equipment cost led to the 'beam waveguide' approach of reflecting energy to and from the subreflector to the feed (now located in a building below the antenna) by means of several mirrors, similar in principle to an optical periscope. This permits all equipment to be on one level, eliminates the need for an elevated room, and yet accomplishes this with acceptably low transmission loss.

FIGURE 3. Horn antenna at Andover, Maine.

FIGURE 4. Antenna at Brewster, Washington, D.C.

FIGURE 5. Antenna at Etam, West Virginia.

FIGURE 6. Antenna in Alaska.

The 500 MHz bandwidth required by Intelsat III (1968) could not be obtained by the maser amplifier (150 MHz bandwidth) used for Intelsat II. A parametric amplifier was successfully designed, but at the cost of increased noise. Increasing antenna diameters from 25.9 to 29.6 m offset the effect of greater noise. A similar approach of increasing diameter to 32 m was later used to offset the additional noise introduced when changing from the original, cryogenically cooled parametric amplifiers to the simpler, more reliable, lower cost Peltier cooled parametric amplifiers. Not only did this change eliminate the need for cryogenics, but it also decreased the high power amplifiers (h.p.a.) power requirement by 35 % between a 32 m and a 25.9 m antenna, an important advantage in these days of high fuel costs.

While mechanical changes in antenna design were being made to reduce cost and improve accessibility, the electrical design was also improved to increase G/T (the ratio of antenna gain G to system noise temperature T is a basic measure of system performance). Aperture efficiencies were improved from 50–60 to 60–70 % by shaping the reflector surface to maximize the area illuminated by the subreflector and thereby increase gain. The spillover past the edge of the antenna was minimized as well, to reduce noise.

The costs of two dozen standard A stations, ordered from 1967–83, were studied to determine trends in pricing. The prices were for 'turnkey' contracts, including antenna, foundation, g.c.e., h.p.as, and control console. The extent of auxiliary equipment (standby power, microwave terminal and multiplex) vary among stations, but some interesting conclusions can be drawn.

The number of stations for which contracts were issued in the indicated time periods is shown in table 1, together with the range of 1967 selling prices and their equivalent cost now. It should be noted that three stations are not included, since their costs are significantly disproportionate to the others. The stations listed in table 1 were built by four contractors.

TABLE 1. NUMBER OF STATIONS FOR WHICH CONTRACTS WERE ISSUED
TOGETHER WITH THEIR COST

period	number of stations	price in 1967 millions of U.S. dollars	equivalent cost now millions of U.S. dollars
late 1960s	6	5–6.5	5.5–6.5
1970s	9	2–3.8	4–5
early 1980s	5	2.5–3	6–7.2

It appears that the relatively high price of the first generation of stations may have been due to writing off development costs, plus the higher costs involved in entering a new field. Once past this phase, the real costs in the 1970s and 1980s show a marked decrease.

Of the operating costs for the large Comsat stations, the largest element (about 50 %) is for personnel, and the next largest (20 %) is for power. The power costs have risen by a factor of about three in the last decade.

(c) High power amplifiers

A unique aspect of satellite communications, as distinct from terrestrial radio relay systems, was the need for multiple access to allow the carriers transmitted from a number of Earth stations, each on a different frequency, to pass through a single satellite amplifier. At the Earth stations, a similar situation arose because several carriers may be transmitted from a single station. To minimize intermodulation among the various carriers, the Earth station power amplifier rating had to be approximately five times more powerful than the sum of the output of these carriers. For the larger Earth stations, this required a h.p.a. of 8 kW to cover the entire 500 MHz bandwidth. Such tubes had never been built at 6 GHz. Therefore, Comsat initiated a development programme to learn how a 2000 h (three month) life might be obtained. The results were successful and later production tubes typically operated for about three years (25 000 h). However, these tubes are expensive and require liquid cooling, which is an operational nuisance. To ameliorate these problems, there has been a trend to use several lower powered h.p.as rather than a single 8–12 kW amplifier, and to use air cooled tubes that are available up to 3 kW. While these power levels are far too high for solid state ampliers, the output tube gains are sufficient to permit all low level stages to be solid state, thus simplifying power supplies, lowering costs and increasing lifetime.

In the large U.S. stations, on-the-air availability is about 99.995 % on average, but of the corresponding 25 min unavailability per year, the major cause is due to problems with the h.p.as. The second largest contributor to unavailability is due to operator error, which arises from the constant reconfiguration required in these large stations that communicate with dozens of other stations.

(d) Low noise amplifiers

The earliest stations, in Andover, Goonhilly, Pleumeur-Bodou and Raisting used masers that had an extremely low noise temperature of 4 K but required operation at liquid helium temperature, with liquid stored in a vacuum insulated flask and periodically topped up to replace the liquid lost through boiling. Later, cryogenic refrigerators were designed by the A. D. Little Co. With the introduction of Intelsat II satellites, Airborne Instruments Lab. developed a maser with a 150 MHz bandwidth (in the 4000 MHz region), but for the 500 MHz requirement of Intelsat III in 1968, it did not appear possible to extend further the maser bandwidth. A 500 MHz parametric amplifier was developed for Comsat by A.I.L. with a noise temperature of 15 K. The increase in noise was offset by adding a 183 cm skirt to our 25.9 m antennas.

These parametric amplifiers operated with gaseous rather than liquid helium at approximately 15 K, which resulted in a simpler, more reliable refrigerator than for the masers. This still required care and periodic overhaul. With the development of better diodes and higher frequency parametric amplifier oscillators, it was found possible to use electrical Peltier cooling and dispense with the mechanical refrigerator completely. Here again, this operational gain was bought at the expense of more noise, about 35 K. This was offset by increasing the antenna diameter to 32 m.

Spurred on by the need for lower cost amplifiers for small Earth stations, intensive development has resulted in gallium arsenide field effect transistors with noise temperatures of about 75 K at 4 GHz. These eliminate the need for a pump oscillator, operate at a low voltage and have a long life.

All of the noise temperatures quoted pertain to the low noise amplifier itself. When the noise contributions due to sky noise, filters, switches and waveguides are included, 45 K is typically added.

(e) Multiplex systems

Initially, Intelsat used F.D.M./F.D.M.A. because F.D.M. was the universal modulation method for cable and microwave relay. It was recognized that passing multiple carriers through a satellite travelling wave tube would provide only half the total capacity possible with T.D.M.A. In 1965, when T.D.M.A. was in its early stages of commercial operation, Intelsat concluded that it would not be desirable to introduce additional untried technology into the system if conventional multiplex could be used, even though the efficiency was decreased. However, the future need for this more efficient system was apparent. In 1965, experiments were started that tested T.D.M.A. This work has continued, and T.D.M.A. will become part of the Intelsat system in 1984. T.D.M.A. has been in use in both the Canadian and United States domestic systems for a number of years.

Intelsat is investigating 32 kbit s⁻¹ digital modulation, half the present rate. Digital speech interpolation may further double or triple the number of possible voice channels by using the gaps in a users speech, as well as the larger gaps that occur when listening to the oppcsite party, to interpose pulses from other channels. In the future, perhaps 16 kbit s⁻¹ codecs might be built with toll quality. Experiments between the Federal Republic of Germany and the United States are under way using S.S.B./A.M., which has the potential of providing five to ten times the channel capacity per transponder, compared to current methods. This has become feasible because of the availability of compandors on a chip and lower speech power per speaker. The latter has come about because of the higher quality of long distance circuits.

(f) Small Earth stations

Intelsat V and VI satellites, with K-band e.r.p. from 41 to 44 dBW will permit the use of standard E stations. The smallest of these, E-1, has a 3.5 m antenna and a 64 kbit s⁻¹ capacity. This antenna can be readily placed on an office building or plant, provide one or two voice channels, or one high speed data channel. The space segment can be leased full time, part time, or for occasional use. Therefore, this should open up entirely new domestic and international services, even though political problems may arise concerning ownership of such stations and space segment access.

3. Fixed domestic service stations

The success of international systems in the 1960s gave rise to the domestic systems of the 1970s in Canada, the United States and Indonesia, plus the leasing of Intelsat transponders or portions of transponders to countries for their domestic needs. At present, 23 countries are leasing a total of 41 transponders. By the second half of the decade, about ten new domestic satellite systems will be launched. Because the antenna coverage pattern is usually less than the pattern required for international use, domestic satellites typically produce about ten times the e.r.p. of an international global beam, which permits the efficient use of 5–10 m antennas. However, as in international service, large antennas are more cost effective for large traffic cross sections. For example, in A.T.&Ts new Telstar 3 system, composed of three domestic satellites using companded single sideband transmission, the transponder telephone channel capacities are 7800 for 30 m antenna diameter with a 2° or 4° satellite spacing; 4200 for a 12 m antenna with 2° spacing, or 6000 for 4° spacing with the same diameter.

Even with a difference of 1800 channels, as at 4°, at an average charge of $750 per month for a leased line, the use of the larger antenna can produce an increase of $1 350 000 per month. With 24 transponders per satellite, the larger antenna is clearly more cost effective. Further, the new F.C.C. requirement for 2° spacing between satellites along the orbital arc, instituted to accommodate the growing demand for new slots, will favour the use of large antennas. This could increase satellite capacity by 86 % over the 12 m capacity for this particular case. For light-traffic stations, the smaller antenna is to be preferred, since the new generations of domestic satellites have more power than their predecessors. To further increase channel capacity, they are operating in both the C- and K-bands. Also, more efficient modulation methods have been introduced, including 32 kbit s⁻¹ digital transmission in place of the standard 64 kbit s⁻¹ rate, digital speech interpolation to further double the capacity, and companded single sideband, so that a 24 transponder Delta class satellite, with a 30 m Earth station antenna, could carry over 90000 two-way telephone circuits.

A 2.4 m antenna used for early domestic system tests is shown as it would be used operationally, in the parking lot of an industrial plant (see figure 7).

FIGURE 7. Business services antenna.

At present, the United States domestic satellite networks include about 500 transmit–receive Earth stations, with antenna diameters ranging from 1.2 to 30 m. The traffic per station ranges from 1200 bit s^{-1} to thousands of voice channels. A.T.&T. assigns as many as 28000 message telephone circuits to satellite facilities. In addition, the Alaskan domestic system has approximately 200 transmit–receive stations, which use, for the most part, 5 m antennas, one or two s.c.p.c. channels and a television receive channel. About 10000 c.a.t.v. receive-only stations, with 4–5 m antennas, are used to feed programs into cable systems. Finally, 400000–500000 C-band receive-only stations of 2.5–5 m diameter are used for television reception by individuals located in places not served by u.h.f. or v.h.f. television stations, or by those who desire a larger variety of programmes.

4. DIRECT BROADCAST STATIONS

In 1977, the Broadcast Satellite World Administrative Radio Conference proposed a worldwide plan at K-band for the direct broadcast satellite service, assigning every country to one or more longitudinal slots and a number of television channels, usually five, but extending to some dozens of channels for the larger countries. This was a quixotic approach, based on a complete absence of experience with direct broadcast satellites. Events since then have proved that this approach has serious problems. Twenty Arab countries joined together in a common satellite for telephony and direct broadcast. Since they had been assigned a variety of orbital slots, there was no slot common to all the countries involved. This predicament forced them to go to the 2000 MHz region for the broadcast transponders.

The slots assigned to China included 35 separate beams from several slots to cover the country. Since they preferred to initiate their system with a less ambitious programme, one covering the country with only two beams from a single satellite, the resulting patterns did not match the preassigned patterns of neighbouring countries. Canada and Australia wanted their satellites to be used for fixed service as well as for broadcast service, a combination not envisioned by the plan. Finally, Earth station technology at 12 GHz, and television technology have generated significant improvements since 1977. North and South America did not accept the 1977 proposal and, in 1983, chose their own assignments, which resulted in a higher number of channels per country than would have been obtained in 1977. One analysis estimates that the use of the orbital arc was nearly four times as great under the 1982 plan as for the 1977 one.

In the United States, numerous satellites are under construction. This demonstration of active interest has spurred manufacturers to step up development efforts on 12 GHz home receivers. Antenna efficiencies have increased from 50–55 to 65–70 %. Improved gallium arsenide field effect transistor devices now provide noise figures of 2.5–3.5 dB, compared to the 6 dB in 1977. Most important, projected home receivers, including antenna, outdoor and indoor electronics with video and audio deciphering circuits and an individually addressable

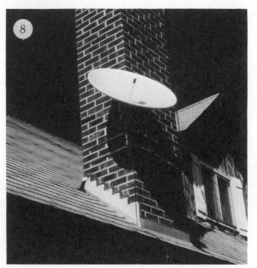

FIGURE 8. Direct broadcast television services.

feature, are anticipated to cost $300 if provided in large quantities. While Canada is already using lower powered satellites, with home receiver antennas of 1.2–1.8 m diameter, the Satellite Television Corporation (S.T.C.) satellites, due to come into operation in early 1986, will be much higher powered and operate with 0.75 m antennas (see figure 8). To cover two time zones in the U.S.A. (to serve about half the United States) and have an edge of coverage e.r.p. of 53 dBW (0.2 MW), 200 W travelling wave tube will be used, manufactured by C.S.F. and Telefunken. The signal will be enciphered, and each home will be individually addressable via satellite. S.T.Cs early entry system will start operation with a slightly lower power beam covering the area from Boston to Washington, D.C. By the end of this decade, five to ten million home receivers may be in operation. Future trends will be towards transmission of higher definition television.

5. MARITIME SERVICE

Maritime service started in 1976 with the successful launches of three Marisat satellites, one over each of the major oceans. The shipboard antennas were 1.2 m in diameter and automatically steerable to keep pointing at the stellite, regardless of the heading of the ship. The

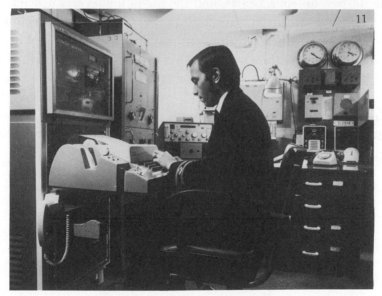

FIGURE 9. A maritime installation.

FIGURE 10. A ship antenna.

FIGURE 11. Below-deck equipment.

ship station could provide one voice and one teleprinter channel. By 1982, these circuits were converted to automatic dialling. They were so successful that over 2000 commercial ships are now so equipped. By increasing the antenna efficiency and lowering receiver noise, it is now possible to use antennas of 0.9 m diameter. Furthermore, even in a period of inflation, the cost per ship station has dropped from \$50 000–60 000 to \$35 000–40 000. Whereas Marisat had a capacity of approximately ten voice channels and 44 teleprinter channels, the next generation of Inmarsat satellites will have a capacity of about 120 voice channels and 300 teleprinter channels. Future trends appear to be in the direction of further decreasing the cost per channel and the cost of ship stations. One approach is to investigate digital transmissions at lower than the present 64 kbit s^{-1} rate to reduce bandwidth and thus the corresponding satellite power or ship station antennas, or both. There is also a search for a much lower cost ship station devoted solely to teleprinter operation. Fifty-six kilobit per second transmission service is now provided from ship to shore, and 1 Mbit s^{-1} service is being considered. Over the years, several proposals have advocated illuminating an ocean with a number of beams, thus permitting use of the smaller ship antennas and frequency re-use, thereby increasing channel capacity.

Figures 9–11 show a maritime antenna installation in a radome, a view of the 1.2 m antenna, and the below-deck installation, respectively.

6. Conclusions

International satellite communications, in less than two decades, have expanded from 60 circuits in 1965 to 33 000 circuits today, and from four stations to 650. The Intelsat system now offers a variety of services designed for Earth stations with antennas from 3.5 to 30 m in diameter. The thrust of Earth station design is to simplify equipment and reduce cost per circuit, even at the expense of more complex satellites, including on-board processing, switching between multiple narrow antenna beams, and the use of higher frequencies.

Domestic and maritime satellite systems are heading in the same direction, towards higher powered, higher capacity satellites and simpler, less expensive Earth or ship stations. This will tend to move the complexity of the satellites, using the techniques described above.

Finally, in the direct broadcast systems of the 1980s, there may be a trend toward higher definition television. Since the existence of this service depends on low cost stations, the industry is looking toward the design of receivers and decoders using circuits on a chip and enabling production quantities in the millions to achieve Earth stations at low cost. This will provide programmes to every home in the nation, regardless of location.

Phil. Trans. R. Soc. Lond. A **312**, 67–73 (1984)   [67]
Printed in Great Britain

Navigation satellites: their future potential

BY W. E. RAMSEY

U.S. Navy, The Pentagon, Room 4C668, Washington, D.C., 20301, U.S.A.

The United States Department of Defense is developing a new generation of navigation satellites known as the Navstar Global Positioning System or GPS. When the full system of 18 satellites is deployed in the late 1980s, highly accurate information on position, velocity, and time will be available continuously to users anywhere in the world. This capability has already been demonstrated by the existing constellation of five prototype satellites.

The original impetus for Navstar was the need for highly accurate positioning information by military aircraft, ships, and ground units. But Navstar also has potential for a variety of civilian uses, which include precision navigation, surveying and accurate time transfer. Moreover, the projected sharp decline in the cost of GPS user equipment will make the system available to a wide class of users.

INTRODUCTION

Improved navigation upon the Earth's surface has been one important benefit to man's venture into space. First launched in the 1960s, navigation satellites now provide accurate navigation information to thousands of ships and other users. A second generation of satellites is now under development by the U.S. Department of Defense. When fully deployed in the 1980s, these new satellites will be a true global positioning system that will provide continuous, highly accurate position, time, and velocity data to users worldwide. The potential for applications of this capability is immense, as the user community is just beginning to realize. This paper describes the existing and planned navigation satellites and then discusses their potential applications.

TODAY: TRANSIT

The concept for the first navigation satellite dates back to the launch of Sputnik I in 1957. While monitoring the famous 'beeps' of the passing satellite, scientists at the Applied Physics Laboratory of Johns Hopkins University noticed that their frequency shifted as the satellite traversed the field of view. Later, they demonstrated that all orbital parameters of the satellite could be determined from this doppler shift. Similarly, if the orbit of the satellite was known, a plot of the doppler shift against time could be used to determine the receiver's position on Earth.

The U.S. Navy was receptive to development of a satellite navigation system because of the need to supply submarines with precise navigation fixes worldwide. The first prototype satellite, launched in April 1960, demonstrated the potential of navigation satellites to meet the Navy's requirements. Continuous operation of navigation satellites began on 12 January 1965. The system, called Transit, represented the first routine use of space technology in direct support of the fleet.

Six operational Transit satellites are now orbiting the Earth in polar orbits at an altitude of 600 nautical miles (1111 km). The system provides a 2σ accuracy of about 460 m when a satellite is in the field of view, which occurs at intervals of 30–120 min. Over 10 000 Transit receivers are now in use, of which about 90 % are operated by civilian users. The rapid growth in Transit users has been accompanied by a steady drop in the cost of user equipment (see figure 1), making satellite navigation economically attractive for more and more users.

The principal commercial use of Transit has been for routine navigation by ocean-going vessels. Transit is also opening new doors in oceanography and offshore oil exploitation by enabling ships to establish a position and return to within a few hundred metres of it.

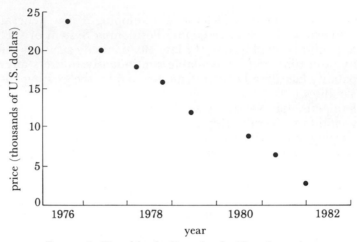

FIGURE 1. Trend in the list price for Transit receivers.

TOMORROW: NAVSTAR

Although it represents a major advance in navigation, Transit is not a true global positioning system. Its coverage is intermittent; it gives position information only in two dimensions; its accuracy is insufficient for many potential applications; and its performance is degraded if the user is moving rapidly, as would be so for aircraft in flight. To overcome these limitations, the U.S. Department of Defense is developing a second generation navigation satellite system, known as the Navstar Global Positioning System (GPS).

How Navstar works

The basic concept for Navstar differs considerably from that of Transit. The latter uses the doppler shift of the signal from a single satellite, measured over several minutes, to estimate the user's position of the surface of the Earth. Navstar uses measurements of signals from multiple satellites to establish the user's position in three dimensions (including altitude), the user's velocity and the exact time. In theory, the user's set sould determine its three position coordinates from only three satellites, but that would require the set to have an extremely accurate, expensive clock. Instead, Navstar employes signals from four satellites, which allows the set to function with a small, inexpensive quartz crystal oscillator.

Given precisely timed pulses from four satellites, the user set automatically solves four simultaneous equations for its three position coordinates and the error in its clock (see figure 2). A similar system of equations that use doppler shift measurements from the same four satellites yields the user's velocity (both magnitude and heading). This procedure gives three-dimensional position worldwide with a median error of 16 m (over ten times more accurate than Transit), three-dimensional velocity to within 0.03 m s^{-1}, and time to within 50 ns.

The space segment of Navstar will consist of 18 satellites arranged in six inclined orbits with an orbital period of 12 h at an altitude of 11 000 nautical miles (20 372 km). Six prototype satellites are currently in orbit, providing one to four hours coverage per day worldwide. A full constellation of 18 production satellites will be in orbit in 1988. (The first production satellites will be launched from the Space Shuttle in 1986.) The ground control segment of Navstar will also be fully operational in 1988, and user equipment will be available from several manufacturers.

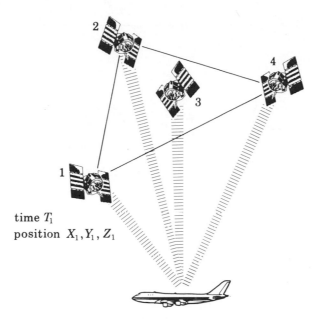

FIGURE 2. Determination of the user's position and the time from four Navstar satellites. The distance from the user to each satellite can be computed from

$$\{(X_i-X_u)^2+(Y_i-Y_u)^2+(Z_i-Z_u)^2\}^{\frac{1}{2}} = c \quad (T_i-T_u-E),$$

where c is the speed of light and E is the error in the user's clock; the resulting four equations can be solved for the four unknowns X_u, Y_u, Z_u, and E.

The Department of Defense is fully committed to Navstar. The three armed services plan to purchase over 20 000 user sets to equip most ships and aircraft, many vehicles, and some man-portable units. The total cost to develop and produce the system and launch it into space will be over two billion dollars. In addition, the Department of Defense plans to spend over one billion dollars on development and production of user equipment.

Test experience

A variety of controlled and operational tests have been conducted with the six prototype satellites and developmental user sets. In general, performance has met or exceeded specifications for all sources of error, including clock errors, ionospheric propagation, multipath effects, and receiver noise. Figure 3 shows the result of a controlled test of a five-channel user set carried by a C-141 aircraft at the Yuma Proving Grounds on 4 February 1983. A laser ranging system was used to determine the actual position of the aircraft and the error in the

Navstar position. The horizontal error averaged between 8 and 10 m, and the total error averaged between 15 and 20 m.

Another controlled test was made on 10 October 1983 with a low-cost, single-channel commercial user set rather than the five-channel set. The results are shown in figure 4. The figure also illustrates how the geometric arrangement of the satellites affects accuracy. For the first portion of the test, before about 13h23, the set was tracking four satellites that were bunched relatively close together. The error in position ranged up to 60 m during this period.

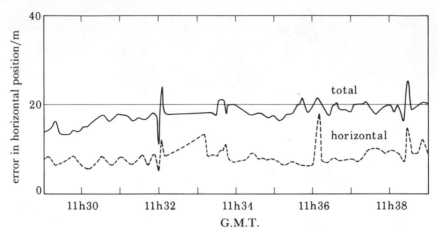

FIGURE 3. Accuracy of the Navstar system in test with a five-channel user set in a C-141 aircraft.

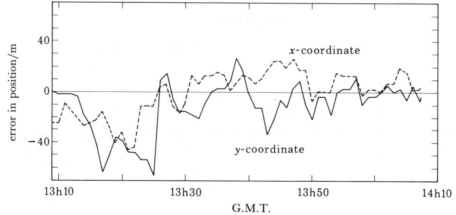

FIGURE 4. Accuracy of the Navstar system with a low-cost user set.

At 13h23 another satellite entered the field of view and increased the geometric dispersion. As the data shows, the accuracy then improved to better than 30 m for the remainder of the test.

A variety of operational tests have also been conducted. In one series of tests tactical aircraft used Navstar to rendezvous with an airborne tanker. The maximum horizontal error during these tests was 9 m. Equally impressive was the recent flight of a Sabreliner business jet from Cedar Rapids, Iowa, to Paris, by navigating solely with Navstar. The flight was made in five segments, timed to match the availability of the existing Navstar satellites. After touchdown at Le Bourget airport, the Sabreliner crew used Navstar to taxi to a targeted parking location without visual cues from ground personnel. The plane stopped just 7.73 m from the pre-designated spot.

Military uses

The Department of Defense developed Navstar, of course, primarily for its military applications. Navstar will be the primary navigation system for military aircraft and ships, allowing the Department of Defense to cut back or terminate operations of existing navigation systems. The most visible action will be the phasing out of Transit, which is scheduled to cease operations in 1992. The Department of Defense also plans to discontinue, in whole or in part, the use of Loran, Omega and Tacan.

The accuracy of Navstar will improve the efficiency of many operations and enable units to conduct some operations that were not possible before Navstar. Military aircraft will be able to land at any airfield, whether or not an instrument landing system is available, under virtually any weather conditions. Operations requiring coordination in position and time, such as the aerial rendezvous for refuelling mentioned above, will be more practical. Coordinated fleet air defence is another example. The lack of accurate position data is currently a serious limitation on coordinated operations by air-defense ships. Navstar will solve this problem, and will also improve the management of air-defence aircraft.

Navstar will greatly improve the efficiency of minesweeping operations. A minesweeper that knows exactly where it is and where it has been will be able to sweep a given area more quickly. Also, the area that must be swept will be smaller because ships with Navstar can navigate a narrower channel. A variety of other military operations will benefit from Navstar.

Potential civilian uses

The Department of Defense does not plan to actively market Navstar in the civil sector; the system's capabilities will have to speak for themselves. But the department will do everything it reasonably can to make the revolutionary capabilities of Navstar available to the widest possible base of civilian users. No doubt Navstar will have a variety of civilian applications, including innovative uses that are not yet envisioned.

Perhaps the most obvious civilian use of the basic Navstar system is for navigation. Navstar would be ideal for cross-country and trans-oceanic navigation by commercial aircraft. Undoubtedly it could have prevented the Korean airliner incident, where the aeroplane flew over Soviet airspace and was shot down in September 1983, and similar instances of navigation errors. Moreover, the relatively low cost of user equipment should make Navstar available to private as well as commercial aircraft. Based on the experience with user equipment for Transit (see figure 1), a basic Navstar user set should cost as little as $500 by the year 2000. Even more complex multichannel sets with computer software for route planning, multiple waypoints, etc. will probably cost just a few thousand dollars.

Maritime navigation is another obvious application for Navstar, not only on oceanic voyages but also in coastal and inland waterways and in approaches to harbours. Navstar would also be a boon to oceanography and to exploration for offshore resources. Transit has already been a major help to these activities. The greater accuracy and continuous availability of Navstar will enable ships, in effect, to draw an X on the ocean to return quickly to precisely the same spot.

Navigation for land vehicles – trucks, cars, and taxis – is also a promising application for

Navstar, particularly if it were integrated with a map display system to show the position of the vehicle. U.S. and Japanese automobile manufacturers are already experimenting with such a system. In the future, a voice synthesizer might even be added to inform the driver of present location, of streets ahead, or a recommended turn.

The application of Navstar to navigation extends into space. The U.S. National Aeronautics and Space Agency is planning to include Navstar equipment on the Space Shuttle to provide navigation information on orbit and during re-entry and landing. Meanwhile, down on Earth, Navstar should provide information on position to individual consumers: private aircraft, pleasure boats, recreational vehicles, even hikers, campers, and hunters.

Other applications of Navstar will depend on additions and refinements to the basic system. One refinement, known as differential positioning, could play an important role in certain civilian applications. Some error sources in Navstar, such as uncertainty in propagation through the ionosphere, are nearly constant over a fairly wide area. A calibrated receiver located at a fixed, surveyed site can determine corrections for these errors and transmit them to local users. With these error corrections the user equipment could achieve an accuracy of about five metres.

With differential positioning, Navstar could be used at thousands of airports around the world to extend the period of safe operation even under poor weather conditions. Differential positioning would not replace instrument landing systems (i.l.s.) at major airports. (Microwave landing systems, the next generation of i.l.s., will be able to guide an aircraft on to the runway automatically.) But differential positioning would be a relatively inexpensive way to improve safety and increase operation hours for the many airports and aircraft not equipped with a microwave landing system. The modest cost projected for Navstar user equipment should allow even general aviation to take advantage of Navstar's capabilities.

'Report-back' is another capability that could be combined with Navstar to produce a system with considerable potential. The idea would be for aircraft, ships, or vehicles to automatically report their position, as indicated by their Navstar user equipment, to a central dispatcher. Tracking trans-oceanic air traffic would be one such application, conceivably by using existing h.f. communications circuits for reporting back. For other applications of report-back, new communications circuits may be necessary.

Static positioning with great accuracy is another important application possible with refinements to Navstar. The first step to greater accuracy is differential positioning. By averaging the resulting readings over time, extreme accuracies can be achieved. For example, Navstar survey equipment has already demonstrated relative positioning accuracy better than one centimetre within one hour on site. Such accuracy could be achieved within tens of kilometres of the reference point. Surveying equipment incorporating Navstar could vastly improve both the accuracy and efficiency of many survey tasks.

Navstar can facilitate accurate transfer of time as well as distance. Its 100 ns accuracy is even superior to that of the portable atomic clock currently used for highly accurate synchronization between distance sources. Navstar will provide an inexpensive way to synchronize computer networks, data encryption systems, communication networks and other navigation systems.

Summary

Navigation satellites were one of the first practical applications of space systems. The second generation of navigation satellites now under development will provide a truly global positioning system, which will provide opportunities for many new and important applications; some of these have been mentioned. No doubt other innovative uses will emerge when Navstar is fully developed and its capabilities and costs become known to potential users.

Discussion

J. R. PAGE (*Kingsland Road, Alton, Hampshire GU34 1LA, U.K.*). Given the potential accuracy of the GPS navigation system, will this lead to the replacement of inertial navigation systems for terminal guidance? Should the military conceive a need for a new navigation system in the future, would the American government guarantee the continuation of the system for civilian uses?

W. E. RAMSEY. It is not intended that the Navstar Global Positioning System (GPS) will replace inertial navigation systems. Instead, it is envisioned that GPS will be used to update inertials and that inertials will be employed as the primary navigation system for terminal guidance.

The current U.S. Department of Defense policy provides for worldwide access to GPS Standard Positioning Service signals at an accuracy level of 100 m by the time the system becomes operational in 1988. Considering the large military investment in GPS and the system's potential for improved accuracy through differential applications, it seems likely that GPS will remain operational well into the twenty-first century. Speculation about development of a new navigation system appears premature at this time.

Phil. Trans. R. Soc. Lond. A **312**, 75–81 (1984)   [75]

Printed in Great Britain

Space: the finance sector

By J. P. MacArthur

Kleinwort, Benson Limited, 20 Fenchurch Street, London EC3M 3BY, U.K.

The paper will deal with the relative importance of finance. Analogies will be drawn with the financing of nuclear power and North Sea oil. The entities – public and private – involved in financing will be summarized including joint ventures, military funding and the impact of privatization. The various types of space expenditures to be financed will be reviewed including ground facilities, launch vehicles, satellites and space stations. The availability and appropriateness of different types of finance will be considered including defence and P.T.T. budgets, resources of Broadcasting Authorities, finance of equipment suppliers and export–import credit. In addition, funds available from national and international capital markets will be considered. The impact of risk and whether it can be insured or hedged will be examined together with the relevance of risk analysis to the availability of private sector funding. In this context the influence of interest rate changes, inflation rates and currency parities will be considered.

Introduction

I believe that finance for space is as crucial a problem as any of the other technical, political, or scientific issues that are discussed in the other papers. The sums involved are so huge and the competing claims for scarce capital resources so many that it is no exaggeration to claim that in the absence of medium to long-term commercial prospects, many technically feasible developments will never take place.

This conclusion is reinforced by the growing acceptance, at least in the mixed economies, that financial resources are ultimately homogeneous. In both national and international economies it is now widely recognized that the conventional distinction between public finance and private finance may be misleading when applied to the overall economic impact of funding major developments. Clearly public rather than private finance may be necessary in certain circumstances for non-economic reasons, but with the important exception of military requirements the acid test is increasingly the likelihood of economic returns.

The reasons for this important change are complex, but perhaps the single most important one is political. Broadly speaking there has, in recent years, been a significant political shift in the West, which has brought into power governments with a practical or philosophical will to control or, if possible, reduce the relative share of the public sector in both the G.D.P. as a flow, and the assets and liabilities of a given economy as a stock. Indeed, in some cases it has been the avowed intention of opposition politicians to adopt such policies that has secured their election; this may in retrospect come to be seen as one of the most profound political changes of the post-war era. The simple proposition underlying this complex change has two essential elements: the first is that what is spent, for example, on space represents a reduction on what can be spent elsewhere; and the second, that publicly-funded spending tends to be bureaucratic and less likely to achieve an economic return than privately funded spending.

What this means in practice varies from country to country and is dependent to an important extent on the influence of the military establishment in resisting the incursions of budget-cutters, on the one hand, and the willingness and ability of private sector companies and mechanisms to pick up the risk and financing burden, on the other.

Relevant financing analogies

Two relevant analogies also serve to highlight the essential differences to which I have already referred. For nuclear power and its peaceful applications (and here I am specifically ignoring the current attempts to recreate a link between military and civil applications), there is no doubt that the extraordinary military, scientific and economic effort devoted to its development stemmed directly from wartime requirements. Equally, there is little or no dispute that the basis on which the technology was made available for civil uses was not strictly economic, financial or even commercial in the sense of implying a return on the original investment (even if that could have been properly costed). In the U.S.A. much of the technology was effectively passed on for development, at low or no cost to industry, for industrial purposes, and in the U.K., the research establishments in which the corpus of knowledge resided were gradually commercialized and benefits passed on to the private sector either directly or indirectly. Beyond the nuclear powers, however, the costing of commercial spin-off was even more extraordinary in that the unbelievably expensive military technology was passed on at very low apparent cost in its civil applications by means of joint venture or licensing agreements not only to France, Belgium and Holland, but also to Germany, Japan and Italy. It must be observed, however, that at this stage of technology transfer, the risks of commercialization were not perceived as being exceptionally high and the economic benefits, even at a time of what now appears to have been 'give away' oil prices, considerable.

The development of the North Sea provides another interesting analogy in terms of the total investment requirement, the level of risk and the régime imposed by Government. In the present context I believe that we will gain most by concentrating on differences rather than on similarities. Originally the oil sector was characterized by very large companies with a history of making huge speculative investments at high risk in the hope of high rewards. Secondly, the timescale was relatively short, and thirdly the market was more or less assured. Finally, it is important to note that the significant entities in the North Sea had considerable commercial freedom and, in particular, the freedom to move. In other words, when, after the initial phase, the Government started to put pressure on North Sea operating companies and introduced what the latter in general felt were 'unfair' taxation and related régimes, they not only threatened to move their rigs and best people to more hospitable offshore climes, but in some cases actually did so. They had in fact better places to go to and the Government decided to make some significant changes in the more controversial tax regulations. The immediate result has been renewed activity in existing blocks and considerable interest in applying for additional blocks in the recent licensing round.

In practice, however, it is very difficult to draw any precise conclusions from either nuclear power or the North Sea in terms of the commercialization of space. My immediate conclusion is intuitive rather than intellectual and consists of the simple proposition that it is essential to ensure that when it happens the U.K. has a significant involvement. This is entirely compatible, for example, with my view that the U.K. should be involved in the development of nuclear power, but that it is relatively unimportant whether we use advanced gas-cooled or pressurized water reactors, or remain abreast of fast breeder reactor technology by means of joint venture agreements.

I have spent some time on the history of this type of major financing problem because I think there are important lessons to be learned in the context of space. Perhaps the most important of these is that once the initial expenditure has been made it is usually sensible to ensure the widest possible spread of commercial benefits by making the technology available as quickly and as cheaply as possible (subject always, of course, to military and political constraints).

Before leaving the broadly philosophical and policy areas I would like to make an observation about space in the context of financing. Space does, of course, represent a new dimension for most of us. Indeed, there are many who would claim that financial institutions have scarcely come to terms yet with earth and water let alone air and space! This elemental charge is not entirely fair but it does raise an important psychological point that has a bearing on the general reaction to financing problems by bankers and investors. Where the rate of change in technology is high and accelerating it does create particular problems for non-specialists in two specific areas; the first is the absence of any build-up of familiarity with the concepts and practices of the technology and the second is the development of statistical and related data on the basis of which commercial and financial risk can be assessed in relation to both historical and expected returns.

SOME GENERAL CRITERIA FOR THE FINANCING OF SPACE

If one now moves on to consider the current position on space financing in the context of the above comments the problems are clear, but the solutions less to. There are, however, some aspects that can be generally identified, for example:

(i) the massive size of investment involved;
(ii) the political importance of space;
(iii) the necessity or likelihood of military involvement;
(iv) the importance of international cooperation;
(v) the slight information on actual risk and return;
(vi) uncertainty about the level of competition from State or commercial entities.

In terms of aggregate investment globally it is evident that the overwhelming proportion of expenditure so far has been by governments either individually or inter-governmentally in jointly funded projects. Equally, most of the expenditure has been made on projects that have a direct or indirect military or political objective. These are projects that are ultimately financed by the taxpayer voluntarily or involuntarily in various parts of the world, but not generally by the private saver (except indirectly).

There are now, however, important developments as expenditure in space both matures and becomes more diversified. Some governments are anxious for both ideological and budgetary reasons to shift the space funding burden to the private sector to the greatest extent possible. The line is generally drawn at the point where the longer term military or political capacity of the State would be damaged by too great a withdrawal of public sector funding. Other governments with less concern about the growth of the State in either political or financial terms are still concerned about the budgetary consequences, particularly where this may be having an impact on the balance of payments and the exchange rate. The problem may be exacerbated where the principal industrial entities involved are themselves part of the public sector and the effective hiving-off of the financing burden becomes impossible because the ultimate guarantor for any form of capital remains the State. In such a case the only way to spread the risk is by involving either foreign governments or commercial concerns in the projects. Similar considerations may apply where the principal users of space facilities are already in the public sector. Most P.T.Ts, for example, are either directly controlled by Departments of State or are publicly controlled corporations. Many broadcasting authorities are also directly or indirectly dependent on public sector funding. In such circumstances it is very difficult to arrange private sector finance, which does not rely ultimately on the guarantee of the State. These considerations are particularly relevant in a U.K. context where some of the entities involved have already been privatized, for example British Aerospace and Cable and Wireless, and others are in the process thereof, for example British Telecom. Furthermore, the B.B.C. (although there is no present intention to float it on the market) has to finance its proposed d.b.s. investment outside the framework of its licence income revenue and without resort to Treasury guarantee.

THE PRINCIPAL AREAS REQUIRING EXTERNAL CAPITAL

When consideration is given to the types of expenditure involved in commercial applications in space it is often difficult to distinguish clearly the relation between financing and return. Projects are not always discrete, either in their applications or their use of facilities. In the first instance this tends to mean that few space projects lend themselves to a project investment approach. Typically the involvement of private sector finance in space at present results from the major corporations already working as main contractors, deciding to make a risk investment themselves to participate in a particular project for which they are invited to tender. Such decisions in turn require funding by the corporations concerned by means of internally generated funds, additional debt or new equity funding. At this stage lenders or investors may or may not be required to assess the risk of the specific space project depending on its size in relation to the corporation as a whole.

At present most of the major facilities required for commercial applications in space are not privately funded nor does it seem likely that many of them will be bought out of the public sector or privatized in the foreseeable future. Some of these facilities represent massive public investments, which in many cases are subject to overriding military requirements that would in themselves create great difficulties in commercially financed or insured projects. The facilities typically required for the space projects generally regarded as commercially viable now and in the foreseeable future are:

(i)   ground facilities, launch, landing and monitoring;
(ii) launch vehicles, including reusable vehicles;
(iii) satellites, for a variety of applications;
(iv) space laboratories;
(v)   space stations and industrial parks.

Of the above facilities, satellites represent the area in which private sector funding is most highly developed, probably because the technology is well tried, the size of investment is manageable and the direct returns are reasonably predictable. Over the past two years or so there has been a great deal of interest in the possibility of privately funded launch vehicles, but so far no deals have actually been struck. At present I believe that General Dynamics has made a proposal to N.A.S.A. that an Atlas Centaur launch vehicle should be funded by the company at an estimated cost of approximately $50 million. An earlier proposal by McDonnell Douglas to finance a Delta launch vehicle at a cost of approximately $30 million has now apparently been withdrawn as has a proposal to fund commercially a Titan launch vehicle. Even in an area as relatively proven as this with estimates of a market as high as $1 billion by 1990, therefore, it is not easy to finalize commercial financing proposals. For upper stage vehicles there have, however, been two interesting developments involving newly formed companies. The first was initiated by O.S.V., which has raised approximately $3 million of venture capital and is in the process of raising a further $30 million for a vehicle designed to carry larger payloads (2268–3175 kg). Another recently formed Corporation, The Cyprus Company, has been established with the initial objective of modifying the second stage of a Delta rocket so that it can be launched from the Shuttle. This second stage rocket would have a payload capability of 4.54 Mg to 6.8 Mg. In view of the fact that payloads of this mass have not yet been built, this project is likely to be longer term. In a related field, The Cyprus Company are also proposing to provide payload servicing facilities, i.e. final preparations for the payload immediately before launch. They claim to have considerable private financing available for this particular investment.

Perhaps the most disappointing aspect of private sector launch vehicle financing recently has been the apparent failure of all attempts to provide substantial private funding for the Shuttle. By this I mean the funding of one or a variety of trips. Approximately two years ago various efforts were made to put together private financing of the order of $1000 million (now probably of the order of $1200 million) to finance one Shuttle. The risks associated with a single launch were clearly enormous and even proposals to spread the investment over a series of launches did not reduce this materially in the minds of potential investors. It is important, however, to distinguish between the private funding of a total launch and the purchase by commercial entities from N.A.S.A. of capacity on a Shuttle vehicle. For the latter, considerable progress has been made in interesting a variety of industrial firms in taking space. This often represents a considerable investment by the companies concerned. McDonnell Douglas and Johnson and Johnson have an exciting joint venture for the manufacture of pharmaceuticals in space, which includes drugs for haemophilia, a one-shot diabetes cure, a blood-thinning drug and a human growth hormone drug. Clearly the markets for such products are enormous and may justify the investments involved, which over a period are likely to run into hundreds of millions of dollars. A company called Microgravity Research Associates is at present working with N.A.S.A. on the manufacture of gallium arsenide in space. This substance is one that may, according to some claims, eventually replace silicon for advanced computer chips. Both Union Carbide and John Deere are involved in the development of specialized materials manufacture in space and are already experimenting with metallurgical furnaces in these conditions. Fibre optics is another area of relevant technology where space conditions appear, in principle, to solve certain 'stringing' problems and a number of companies are working on this technology. Although most of the examples quoted refer inevitably to U.S. experience, there are in fact several European and other initiatives with launch vehicles, satellite construction and operation and Spacelab activity. So far, however, the funding for these applications has come either from State or quasi-State entities or from the contractors or users themselves, rather than from investors or direct third party project finance.

The whole subject of permanent stations with laboratories, space stations or industrial parks in space is one that has excited a good deal of attention. Theoretically, and to a degree in practice, there appear to be no reasons why such facilities should not be available before the end of the century in substantial volume. What we are concerned about today, however, are the prospects for private sector financing of some or all of these facilities. Fairchild have a highly developed proposal to provide a space platform that would be available on a lease basis for customers with requirements going from quite small to large. The estimated cost of developing the first platform is approximately $200 million and there appears to be considerable industrial and financial interest in the project.

It is clearly no accident that the part of space financing that is most highly developed is that of satellites. The private sector financing of satellites is based on several of the key criteria already touched on, i.e.

(i) lengthy experience;
(ii) reliability;
(iii) good demand;
(iv) adequate investment return;
(v) low technical and financial risk.

When all these aspects are related to the fact that the users and builders are usually powerful entities with pressing needs for these services, for example Broadcasting Authorities, P.T.Ts and Aerospace Corporations, it is clear that a powerful momentum has been developed. Whether this encouraging pattern of public and private investment will continue on quite such an even keel in the future is more debatable because of the possibly giant strides in technology that may follow from some of the developments outlined above. There are also developments on Earth that paradoxically affect investment in space, for example in economic terms some of the fibre optic interactive cable systems now being proposed could cut into the estimated market for d.b.s. to an extent that would render the latter only marginally profitable or wholly unprofitable.

TYPES OF FINANCE AVAILABLE

I have already spoken of the homogeneity of aggregate financial resources and of the sometimes arbitrary distinctions between public and private financing. Nevertheless, it is still useful to summarize very briefly the types of finance that may be available. Apart from general budgetary allocations under a variety of headings, for example scientific research, there are more specific sources emanating from Departments of State or Government. A possible non-exhaustive list is now given:

(a) defence budgets;
(b) P.T.T. budgets;
(c) Broadcasting Authority budgets;
(d) equipment supplier funding;
(e) export–import credit from foreign suppliers;
(f) bank finance guaranteed by State entities;
(g) bank finance based on the project returns;
(h) capital from joint venture partners;
(i) customer down payments;
(j) risk or venture capital.

CONCLUSIONS

In summary, there is no doubt that an adequate volume of capital resources is available in aggregate to meet the estimated financing needs of space. The real problem arises in assessing what type of capital is required and available, and from what existing uses it will be diverted (to the extent that new capital is not created in the short term). Once the funding of new fields of activity such as those in space begins to move into the private sector, different criteria for the evaluation of risk and reward start to apply, and at the very end of the private sector spectrum the non-specialist fund manager will seek to evaluate the risk:reward ratio on a particular space project against other opportunities available to him whether in gilt-edged securities or listed equity investments. In practice, of course, there are specialist funds that will invest solely in equities or solely in unquoted or solely in high technology investments. Over time it is likely that the financing of space will portray a similar profile to that of other mature but growing industries. In other words it will show, across a spectrum of risk and return, a variety of different funding methods appropriate to the particular project, company or industry. Two key factors will determine the evolution of this profile. The first is familiarity and the second uncertainty (risk). As I have already noted, one is seeing the first classical signs of involvement by risk money, most highly-developed for satellites, but beginning to develop, through the aerospace companies, in launch vehicles and space stations. In my opinion, it is highly likely that within 10 or 15 years most major financing institutions will employ staff with the necessary skills to evaluate risk in a wide variety of space applications with a view to providing capital on a lending or equity basis. In this growing business of evaluating risk and capital provision in space there will be the same need to consider currency factors, relative inflation rates and interest rates over a longer term, which are already used in the analysis of other forms of lending or investment whether traditional or high technology. I now end by uttering the ultimate heresy: space, in its financing aspects at least, will become in the longer term like any other business.

Phil. Trans. R. Soc. Lond. A **312**, 83–88 (1984)   [83]
Printed in Great Britain

The Ariane programme

By M. Bignier and J. Vandenkerckhove

European Space Agency, 8–10 Rue Mario Nikis, Paris, France

The Ariane programme was decided upon in 1973. It was made a European concern after a proposal from France.

Ariane 1, launched for the first time on 24 December 1979, was qualified after three successes from four flights. The Ariane 3 version will fly in mid-1984 and Ariane 4 at the beginning of 1986. The main specifications of these programmes will be described. Their economic interest is shown by the number of orders and options already received. A new launcher will be useful before the end of the present century and relevant studies have started; of the engine in particular.

Introduction

The decision to develop Ariane was taken during the spring of 1973 at the Brussels Ministerial Conference after having been sponsored by France as part of a political package deal, which also included Spacelab and Marots.

The objective of the programme was the provision of an independent launch capability, in particular for geostationary telecommunications satellites, for which the launcher was optimized. Europe wanted to be able to launch operational satellites and N.A.S.A. could not guarantee this. Symphony was launched with the proviso that it would only be experimental. It is fair to add that at that time the commercial prospects of becoming competitive on the world market were largely overlooked.

The performance objectives were particularly well selected: 1600 kg in geostationary transfer orbit (g.t.o.) corresponding to satellites of up to 1000 kg, thus anticipating the needs of users a decade in advance.

A brief history of Ariane 1

Ariane 1 was an E.S.A. programme managed by a Programme Board, which decided on funding and objectives and monitored all development and construction activities. However, unlike other E.S.A. programmes, the management was delegated to C.N.E.S., which acted as prime contractor for the development activities.

In turn, C.N.E.S. placed contracts with European industry, in particular (the present arrangement) with: SNIAS, industrial architect (system level) and main contractor for the first and third stages; SEP, main contractor for propulsion (all stages); ERNO, main contractor for second stages; MATRA, main contractor for the equipment bay; and CONTRAVES, main contractor for the shroud. Subcontracts were widely distributed, Ferranti being responsible for the inertial guidance platform and M.S.D.S. for the flight programme.

The development programme was somewhat limited and relied to the maximum possible extent on hardware that had already been developed. Costs amounted to roughly one billion (10^9) dollars, *without any previous stage flights* and with only four qualification flights of the complete vehicle.

Flight history

The first flight, LO-1, occurred in December 1979, with complete success. The second, LO-2, failed in May 1980, as the result of unstable combustion in one of the first-stage motors, which led to a comprehensive programme of testing and to the decision to test-fire all Viking motors for acceptance. The last two flights, LO-3 and LO-4, which were launched respectively in June 1981 and December 1981, were again complete successes, which launched operational E.S.A. satellites, Meteosat and Marecs 1, as well as an Indian experimental satellite, Apple.

Before qualification, which was announced early in 1982, E.S.A. also decided to order seven additional Ariane 1 rockets, the so-called 'promotion series', the first of which, L-5, due to launch E.S.As Marecs 2 and Sirio 1, simultaneously, failed in September 1982, just before reaching orbit; the failure was a result of a gear problem in the turbo-pump of the third stage. After this, not only were the necessary improvements introduced in the gear box, but in addition, the programme was subjected to a thorough review and many other measures were taken. These steps led, in June 1983, to the successful injection of E.S.As E.C.S.-1 together with a small amateur radio satellite, Amsat (which, however, was subsequently struck by the third stage), into orbit by L-6. The last firing, L-7, in October 1983, was a complete success; it put Intelsat V, with a mass of 1834 kg in g.t.o., about 15 % higher than the original performance objective, into orbit.

The next two launches under the promotion series are also Intelsat V satellites, with L-8 now scheduled for February 1984† and L-9 for two months later. The last two Ariane 1 rockets will subsequently be used for C.N.E.Ss Spot and E.S.As Giotto.

The whole programme so far, which has remained reasonably well within the budget (at least below the 120 % mark), has slipped behind schedule by about 18 months in total. It has achieved five successes from seven attempts. This compares favourably with other programmes: three from eight for Atlas Centaur and seven from twelve for Tital III, which subsequently demonstrated 89 % reliability over 54 launches and 94 % reliability over 122 launches, respectively.

Ariane 1 is a three-stage satellite launcher of simple classical design. Its lift-off mass is 210 t and its height 47 m. The first two stages have, respectively, four and one Viking motors, which use storable N_2O_4–UDMH propellants, but the third-stage motor, HM-7, uses cryogenic propellants (liquid hydrogen and oxygen). The launch range is at Kourou, in French Guiana, only 5° N above the Equator, which enables the maximum advantage to be taken of the Earth's rotation for g.t.o.

Improved versions: Ariane 2, 3 and 4

The development of improved versions of Ariane (figure 1) are in progress. The aims are to:

(i) improve cost effectiveness by taking advantage of scale; in particular, it will be possible to launch two PAM-D-sized satellites simultaneously, by means of a special structure, Sylda;

(ii) launch larger and heavier satellites in the future;

(iii) improve reliability by means of additional redundancies; and

(iv) provide more volume for the satellite designers.

† L-8 was successfully launched on 4 March 1984.

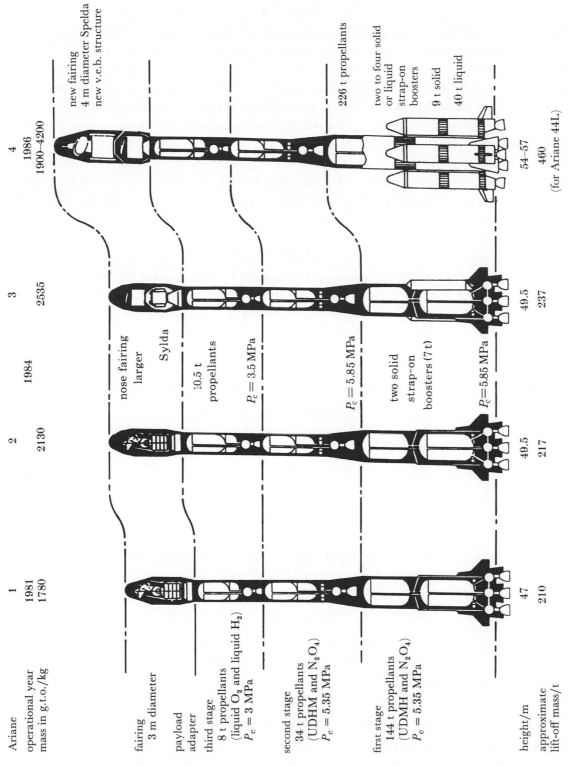

Ariane	1	2	3	4
operational year	1981	1984		1986
mass in g.t.o./kg	1780	2130	2535	1900–4200

fairing
3 m diameter — nose fairing larger; Sylda — new fairing 4 m diameter Spelda; new v.e.b. structure

payload adapter

third stage
8 t propellants
(liquid O_2 and liquid H_2)
$P_c = 3$ MPa — 10.5 t propellants; $P_c = 3.5$ MPa

second stage
34 t propellants
(UDHM and N_2O_4)
$P_c = 5.35$ MPa — $P_c = 5.85$ MPa

226 t propellants

first stage
144 t propellants
(UDMH and N_2O_4)
$P_c = 5.35$ MPa — two solid strap-on boosters (7 t); $P_c = 5.85$ MPa — two to four solid or liquid strap-on boosters; 9 t solid; 40 t liquid

height/m	47	49.5	49.5	54–57
approximate lift-off mass/t	210	217	237	460 (for Ariane 44L)

FIGURE 1. The development of improved versions of Ariane. P_c is the chamber pressure.

Ariane 2, 3

Ariane 2, 3 is a relatively straightforward derivative of Ariane 1. Its core, Ar-2, has somewhat higher performances thanks to the increased combustion pressure of the Viking motors (5.85 MPa instead of 5.35 MPa), and of the HM-7 motor (3.5 MPa against 3 MPa), together with longer third stage tanks (10.5 t of propellant instead of 8 t). The addition of two solid propellant strap-on boosters, manufactured by B.P.D., gives Ariane 3, which has a lift-off mass of 237 t and the capability to inject 2535 kg into g.t.o., a 38 % improvement in performance over Ariane 1. Ariane 3 will be needed to launch E.S.As Olympus.

The first flight of Ariane 3, L-10, is foreseen for June 1984. It will carry two operational communications satellites, probably E.S.As ECS-2 and an American domestic satellite, G-Star 1. Subsequently, recovery of the first stage by parachute for re-use after refurbishment, is also to be attempted (probably for L-13).

Ariane 4

Ariane 4 is a much enlarged derivative of Ariane 3. Its core, Ariane 40, has a longer first stage and its fairing is much enlarged (in particular with a diameter of 4 m instead of 3 m). The performance of Ariane 40 is close to that of Ariane 2, 1900 kg in g.t.o. Greater capability can be obtained by adding liquid or solid strap-on boosters (table 1). The lift-off mass of the later version is 460 t.

TABLE 1. INCREASED PERFORMANCE OF ARIANE 4 BY THE
ADDITION OF STRAP-ON BOOSTERS

booster type	Ariane code number	$10^{-3} \times$ (mass in g.t.o.)/kg
two solid	42P	2.6
four solid	44P	3.0
two liquid	42L	3.2
two liquid and two solid	44PL	3.7
four liquid	44L	4.2

The first launch of Ariane 4, probably the 44PL version, is presently scheduled for early 1986 with three satellites: a Meteosat, Eurostar and Arsène. Ariane 44PL will be capable of launching Intelsat VI.

ELA-2

For launching Ariane 4, which will have a height of up to 57 m, a new launch pad, ELA-2, will be built at the range, which will also handle Ariane 3 rockets. This will allow the annual launch rate to be increased to 12.

TOWARDS COMMERCIALIZATION

Ariane is a competitive launcher.

(i) Its capability meets, almost ideally, the requirements of planned geostationary satellites (up to 1400 kg for Ariane 3 and 2500 kg for Ariane 4), which represent 80 % of the market; of the order of $10 billion until 1990, which represents roughly 200 satellites to launch over six years.

(ii) Its price is substantially lower than that of comparable expendable launchers: Delta, Atlas Centaur and Titan III, as well as some of the Shuttle's upper stage combinations, namely STS/PAM-A and IUS.

For the very important class of 600–800 kg satellites in g.E.o., i.e. half an Ariane 3 or STS/PAM-D or Dz, and with the promotional price of the Shuttle for the period 1985–1988, it is grossly comparable; somewhat lower or higher, depending on the specific case.

(iii) It is available during a period in which there is a shortage of launch capacity.

(iv) It also presents a number of specific advantages, such as high accuracy in placing satellites in orbit (equivalent to a likely increase in satellite lifetime of about 7.5 months), or no extra costs for long payloads; however, the available diameter is smaller than that of the STS.

It is clear that Ariane has introduced real competition in the launcher business; its main competitor during the coming years will be the Shuttle. In the long run, however, this could change; on the one hand because the STS pricing policy will be modified after 1988 (perhaps towards more comprehensive charging) and, on the other, because initiatives are likely to be taken to commercialize unmanned expendable U.S. launch vehicles (e.l.vs). Finally, it is becoming increasingly realized that astronauts are not an asset for the majority of missions, and in the United States some consideration is being given to the development of an advanced expendable launcher, while Japan and China are developing their own capability. The Russians have shown interest, for at least one example (Inmarsat) in selling launch services.

This analysis is supported by the verdict of many users. Beyond the promotion series, firm orders have already been given for the launch of 24 satellites, in addition to which nine options have been taken. Many of them, which add to the first three Intelsat V satellites in the promotion series, are for non-European satellites, often from private U.S. carriers, another three Intelsat Vs, two G-Stars, one Arabsat, one Westar, two Spacenets and two SBTSs.

To commercialize Ariane, a private firm, Arianespace, was created in 1980, with a capital of 179 million French francs, and with the following tasks: sale of launch services; purchase of launchers; performance of launch operation; and maintenance of production and launch capability. Arianespace is the first commercial company to sell satellite launches in the world. The development tasks, however, remain the responsibility of C.N.E.S., which also provides technical support to the firm.

The objective of the programme is to launch at least five or six Arianes per year from 1984 onwards (a majority carrying two satellites), with a high degree of reliability and dependability. This is a major challenge, which requires special measures beyond the development phases. A comprehensive industrialization effort is presently in progress aimed at streamlining the manufacture of the critical hardware. In a number of instances, product improvement efforts are already under consideration, as much of the equipment will continue to be used on Ariane 3 and then Ariane 4.

BEYOND ARIANE 4

While Ariane 4 is expected to meet needs until 1995–2000, plans are already being studied at C.N.E.S. and E.S.A. to develop its successor, emphasis being placed on cost effectiveness and reliability; in particular, the development of a high-thrust cryogenic motor, HM-60, is being proposed to the E.S.A. Member States, since such an engine will be needed irrespective of the final configuration, and its development time is of the order of 12 years.

Discussion

G. M. WEBB (*Commercial Space Technologies Ltd, Hanwell, London, U.K.*). Is not Dr Vanden-kerckhove's expression of E.S.As total reliance on the HM-60 cryogenic rocket engine as 'needed independently of the final configuration' a little incautious in the face of developments in advanced engine design currently taking place in certain British and German companies?

J. VANDENKERCKHOVE. At present only one cryogenic rocket engine, HM-7, which is used on the Ariane third stage, is available in Europe. Its thrust is not sufficient for use on the first or second stage of a large launcher; therefore a higher thrust motor, more advanced than the Viking, is needed. Optimization studies show that the choice of the vehicle configuration does not significantly influence the choice of its thrust level, around 90 t, to inject 12–15 t in l.E.o.

The development of cryogenic HM-60, with this thrust level, is presently being proposed to E.S.A. Member States.

Phil. Trans. R. Soc. Lond. A **312**, 89–102 (1984)   [89]
Printed in Great Britain

The Space Shuttle system

By C. J. Meechan

Strategic Planning Department, Rockwell International, North American Space Operations,
12214 Lakewood Boulevard, Downey, California 90241, U.S.A.

The development and subsequent operation of the Shuttle as America's primary space transportation system is the culmination of several decades of research, technology application and engineering. This vast endeavour, which has enjoyed unprecedented success to date, is one of the great team efforts in our technological history. From the pioneering efforts of Goddard and Tsiolkovsky, the critical contributions of von Karman, the insight of Faget, and the dedicated efforts of a multitude of others, the Shuttle has become a reality.

The initial phases of Shuttle development involved extensive analysis and testing with margins carefully engineered to provide for technical contingencies, and operational results have verified the elegance of the design.

Though its capability to carry cargo has received the most attention, the Shuttle's capabilities in orbit will undoubtedly play a key role in the development of entirely new technologies and associated industries peculiar to the space environment.

This paper will discuss the evolution of the Shuttle system and postulate its future contributions to the industrialization of space.

Introduction

Space vehicles with capabilities like the Shuttle's were conceived long before a technology base was available to make them a practical consideration. Through the successively more sophisticated Mercury, Gemini and Apollo programmes of the 1960s, propulsion technology was greatly advanced and much was learnt about re-entry of spacecraft into the Earth's atmosphere. The technical jump from the 60 000 lb (27 000 kg) Apollo Command Module to the 240 000 lb (109 000 kg) Orbiter, with aeroplane geometry, was impressive in itself. But more importantly, unlike its predecessors, the Orbiter is a 100-mission reusable spacecraft designed for years of reliable, easily maintainable service.

The most versatile spacecraft ever designed, the Space Shuttle, is already expanding our thinking and planning about the future use of space. Along with its 65 000 lb (29 500 kg) payload capacity, the Shuttle provides access to and from space for men and equipment, relatively unlimited payload capability through multiple missions, and unmatched versatility of operations once in orbit. By making space flight a routine event, the Shuttle will transform the way we operate in space.

Shuttle development and testing

The development phase of the Shuttle programme was characterized by extensive tests of all major subsystems and led to the first orbital Shuttle flight in April 1981. I will highlight several areas of interest in design and verification testing and summarize the overall system flight-test results.

FIGURE 1. Two tiles from the Shuttle's thermal protection system.

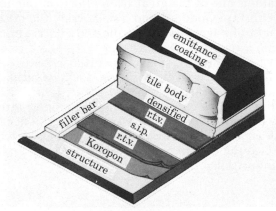

FIGURE 2. The tile fabrication process.

Thermal protection

The Shuttle's thermal protection system includes over 20 000 tiles with a total mass of less than 15 000 lb (6.8 Mg). The tile material is silica fibre with low tensile strength in the 15–30 lb f. in^{-2} (103–206 kPa) range. The basic purpose of the tile design is to provide thermal protection at the bond line of the aluminium structure. That requirement determines the thickness and often the shape of the tile. Therefore, essentially every tile in the system is different (figure 1); almost every one has a different part number and a different set of design loads. Furthermore, there are different coatings on the tiles with different optical properties for thermal control of the vehicle during the various flight phases.

Loads and stresses during nominal orbits, contingency orbits, return to launch sites, and other phases vary with time and flight condition for every tile. These variations required an analysis for each condition of 20 000 individual tile systems; as can be appreciated, it was a major analytical endeavour.

The tile fabrication process is shown in figure 2. The basic structure has a Koropon coating that protects the aluminium. A room-temperature vulcanizer (r.t.v.) is applied to the Koropon so that a cohesive bond is established with the strain isolation pad (s.i.p.). The s.i.p. reduces the loads on the tile caused by structural deflection. The tile is bonded to the s.i.p. with additional r.t.v. A densified layer at the base of the tile distributes the loads uniformly into the tile. Finally, there is black borosilicate emittance coating. (It is interesting to note that, after five flights, the part numbers on Columbia's tiles were still legible (figure 3).)

FIGURE 3. Tile installation.

Beginning with the second Shuttle Orbiter, Challenger, we began to replace some of the tiles on the upper portion of the vehicle with lightweight thermal blankets. The blankets are composed of the same silica material as the tiles, but are sandwiched between an inner and outer quilted blanket. They are designed to reduce fabrication and installation cost, and schedule time, and to further reduce vehicle weight. During Challenger's first orbital flight (STS-6) in April 1983, some blanket degradation, due to either vortex impingement or buffeting on the blunt forward face of the pods housing the orbital manoeuvring system and the reaction control system (figure 4), occurred. Analysis of the problem and flight testing of several blankets in selected areas of the vehicle during the STS-8 mission confirmed their integrity and gave us the confidence to continue the transition to blankets by replacing virtually all of the upper surface tiles on subsequent Orbiters.

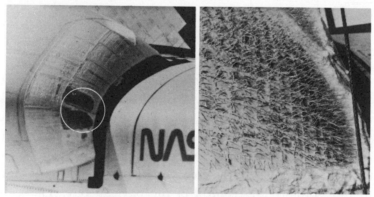

FIGURE 4. Unique aerothermal effects: blanket degradation.

Crew compartment

The crew compartment is an excellent example of a system engineering approach that involves many disciplines. Each mission phase – launch, in orbit, and re-entry to landing – represents a separate set of crew compartment requirements. At the heart of the crew compartment, other than the astronauts, of course, are the computers that form the nerve centre of the system. Today's Shuttle pilots soon become expert computer operators. The rotational hand controllers, rate gyros, and horizontal situation indicators give the impression of a conventional aeroplane cockpit. The selection switches for the autopilot modes are conveniently located on the display panel above the instruments. The three cathode ray tubes (c.r.ts) display anything the crew needs to know for operation: exactly where the vehicle is and where it is going to be. As the Orbiter approaches the landing site, for instance, the c.r.ts display where the vehicle will be in projected 20 s intervals if the present attitude is maintained. In addition, there is a 'heads-up' display so that the pilot can see flight information while looking through the forward window.

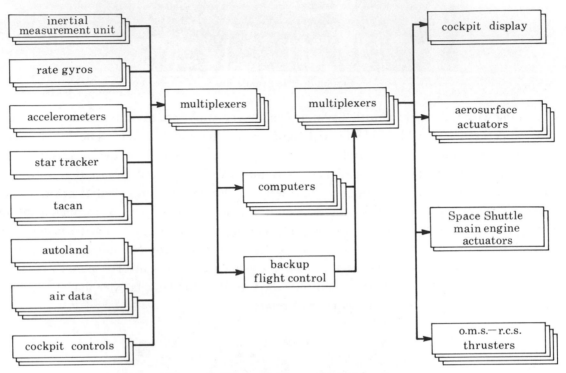

FIGURE 5. The navigation and control system.

Figure 5 is a simplified navigation and control system block diagram. The system redundancy provides a fail-operational/fail-safe system. In addition, a software back-up system constantly follows the primary system so that it can take over if a problem occurs. The back-up system uses separate software developed with a completely different approach so that a generic problem cannot arise in both primary and secondary systems.

The aft station is an integral part of the crew compartment. The c.r.t. and console displays give the astronaut full visibility and control of the payload bay. The astronaut also has a rotational controller to fly the Orbiter from this station. The windows, both above and facing the payload bay, allow rendezvous operations to be effected. The Shuttle's ability to carry a crew of seven is a major step forward in manned space operations (figure 6). In addition to the flight deck, there is a mid-deck for crew habitation and a lower compartment for equipment storage.

Main engines

The reusable main engines are a story in themselves (figure 7). They are very high performance engines, with oxygen pressures up to about 8000 lb f. in^{-2} (55.2 MPa) to feed the preburners. Very little enthalpy is lost; everything is recycled through the main combustion chamber to achieve the required maximum specific impulse. Furthermore, these engines are reusable. In previous space programmes, all engines finished up in the sea. Now that they are returned for inspection, we can refurbish and bring them that much closer to perfection for the next flight. We have never had that opportunity before.

FIGURE 6. The crew compartment.

FIGURE 7. A main engine; it is reusable, lightweight, computer controlled and has a regenerative cycle.

The engines are also computer controlled. The computer is physically mounted on the engine, which is an unusual place to mount a computer, yet the engine performs so smoothly that the *g*-levels have been well below those the computer were designed to withstand. The main propulsion test stand in Mississippi includes the external tank and the actual aft end of the Orbiter vehicle, complete with the large 17 inch (43.2 cm) disconnecters, the engines, controllers, and other elements required for extensive combined system tests. Those tests have been very successful, and have lasted for about 165 ks in total. With engine performance verified, our testing now emphasizes their life and maintainability. There has been little trouble during boost; one of the basic reasons is that the engines have operated so well that the performance traces for them look like the computer plots calculated before flight. The jagged black line in

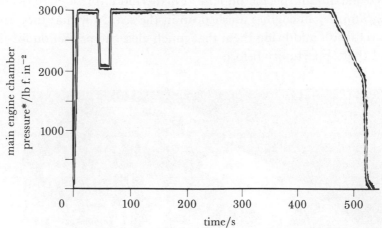

FIGURE 8. Comparison of theoretical main engine performance with actual performance. * 1 lb f.in^{-2} = 6.895 × 10^3 Pa.

FIGURE 9. The external tank.

figure 8 represents the actual preflight computer projections, while the straight line over it is a composite of main engine performance during the first five missions.

External tank

The enormous external tank (e.t.) supplies the fuel used by the Shuttle's main engines (figure 9). An imposing 155 ft (51 m) tall and 28 ft (9 m) wide, the e.t. contains approximately 1.58 million pounds (718 Mg) of usable propellant. The fuel, a combination of liquid hydrogen and liquid oxygen, is consumed by the main engines at a rate of 197 000 lb min^{-1} (90 Mg min^{-1}). After main engine cut off, at approximately eight minutes into flight, the e.t. is jettisoned and breaks up over the Indian Ocean. The e.t. is the only major Shuttle component that is not reusable.

Solid rocket boosters

Designed to be recovered and for use on twenty missions, the bulk of the 149 ft (33 m) long solid rocket boosters (s.r.b.) is the solid rocket motor (figure 10). Built by the Thiokol Corporation, the s.r.bs are the largest solid rockets ever to be flown and the first designed for re-use. A segmented case design affords maximum flexibility in fabrication and ease of transportation and handling.

S.r.b. tests entailed static firings of both a developmental and a qualification nature. Five developmental and four qualification firings were successfully made. In addition, to increase Shuttle performance by 5500 lb (or 2500 kg), a lightweight, filament-wound solid rocket motor is under development and is scheduled to become operational in late 1985.

FIGURE 10. Solid rocket booster test firing.

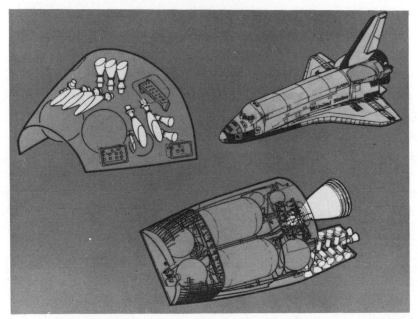

FIGURE 11. Orbiter manoeuvring capability.

Orbiter propulsion

The manoeuvring capability in orbit is made possible by two sets of engines (figure 11), the orbital manoeuvring system (o.m.s.) and the reaction control system (r.c.s.). The two o.m.s. engines provide the thrust for final entry into orbit, orbit circularization, orbit transfer, rendezvous and departure from orbit. Each o.m.s. engine produces 6000 lb (26.7 kN) of thrust. The o.m.s. is housed in two independent pods, located on each side of the Orbiter's aft fuselage, which also house the aft reaction control system.

The overall r.c.s. consists of a series of 44 engines that provide the thrust for attitude changes, manoeuvres (pitch, yaw, roll), and the thrust for small velocity changes along the Orbiter axis (translation manoeuvres) when the orbiter is above 70000 ft (21336 m). The forward r.c.s. module contains 14 primary thrusters and has two vernier thrusters, while the aft r.c.s. has 24 primary and 4 vernier thrusters. The primary engines provide 870 lb (3.87 kN) of thrust and the vernier engines provide 24 lb (107 N) of thrust. The r.c.s. engines are highly redundant; if one is lost, the capability of the system is not.

DEVELOPMENT AND TEST SUMMARY

To summarize, the two principal development results to date are that the system works and it is reusable. Refurbishment requirements are below those expected, which is reducing the turn around time. The consumables have been about 15 % lower than expected, which presages a longer time in orbit. Nine missions have been flown to date; six of these were on one spacecraft. There are at least 94 to go, so there is a long way to go.

FLIGHT ACCOMPLISHMENTS AND RESULTS
STS-1 (12–14 *April* 1981)

The first Shuttle mission was truly remarkable for the first flight of a winged vehicle in a hypersonic environment in which no vehicle had been flown before. The STS-1 mission verified the integrity of the vehicle as all systems met or exceeded test goals.

STS-2 (12–14 *November* 1981)

The second Shuttle mission carried a 30 ft (9.14 m) long Imaging Radar antenna that was very successful in that it revealed sub-surface features of the Earth that had previously gone undetected. In the Sahara Desert, the radar uncovered ancient rivers that have been covered by sand for nearly 5000 years. Because the Sahara is the most arid region of the world, the radar was able to penetrate the sand as deep as 2 to 6 m. Analysis since this discovery has determined that there were flowing rivers in the area during the time of the pyramids, approximately 5000 years ago. Because of the great success of the radar antenna, it is being upgraded and will fly again on a Shuttle mission in summer 1984.

STS-3 (22–30 *March* 1982)

During STS-3 Commander Jack Lousma and Pilot Gordon Fullerton put the Shuttle's Canadian built robot arm through tests with a 400 lb (180 kg) test article. The arm is 50 ft long, (16 m) 15 inches in diameter (40 cm) and has mass 994 lg (452 kg). Four arms are being built by the Canadian government with an investment of approximately $100 million. It is operated in both automatic and manual modes from the Orbiter crew compartment. A mission specialist operates the arm by using dedicated controls, closed circuit television monitors and direct viewing through aft and overhead crew compartment windows.

The arm is a major Shuttle subsystem that will enhance payload servicing and repair, provide support for activities outside the spacecraft, and in the near future will be used to build large structures in space.

STS-4 (27 *June to* 4 *July* 1982)

President Reagan attended the fourth Shuttle landing, which completed its orbital flight tests on 4 July 1982. During his address he compared that landing to the driving of the golden spike that completed the first transcontinental railroad. To round the day off, the President gave consent for the second Shuttle Orbiter, Challenger, to begin its piggyback journey on a 747 to Florida in preparation for its maiden flight the following year.

STS-5 (11–16 *November* 1982)

The fifth Shuttle mission ushered in the operational phase of the Shuttle programme by delivering two communications satellites to orbit. The first satellite to be deployed by the Shuttle is owned by Satellite Business Systems (S.B.S.) of Virginia. S.B.S. is a joint venture of Comsat, I.B.M. and Aetna Life Insurance. This $40 million satellite was the third in a six satellite system that provides computer data, voice, video, and electronic mail services to American firms.

The fifth Shuttle mission also carried a satellite for Telesat of Canada, the Canadian phone company, which is used to relay television and phone messages throughout the North American continent.

STS-6 (4–9 *April* 1983)

During the sixth Shuttle mission we deployed the Tracking Data and Relay satellite, the largest non-military satellite ever built. The satellite was the first of a three-satellite system designed to replace N.A.S.As 20 year old ground-based tracking network.

The activities outside the Shuttle on the sixth Shuttle flight were the first by U.S. astronauts in nine years. They showed the importance of attention to detail: handholds, tethering provisions, and foot restraints, for example, engineered into the payload bay. Many mechanical functions were tested by using special tools designed for use in the absence of gravity. The 3.5 h test outside the Shuttle was a precursor to a capability that will come much more into play as this man–machine system's assignments in orbit evolve.

STS-7 (18–24 *June* 1983)

The five astronaut crew on the seventh Shuttle flight included the first American female astronaut, Sally Ride. The Shuttle's Canadian robot arm was again used, this time to actually deploy and retrieve a German satellite, the Shuttle Pallet Satellite known as SPAS; it was built by the West German firm, M.B.B. It is planned to use SPAS as the basis of a low cost commercial remote sensing platform (SPAR-X).

We received exceptional photography from cameras on the SPAS as it and the Orbiter played a game of 'orbital tag', which demonstrated the Shuttle's rendezvous and docking capability. The SPAS will fly again on the next Shuttle mission (STS-11 in January 1984). During STS-11, an astronaut with a backpack manoeuvring device will use the SPAS to practice docking procedures without a tether. STS-11 is a pathfinder mission that will lay the groundwork for the Shuttle's first satellite repair mission scheduled for April 1984.

STS-8 (30 *August to* 5 *September* 1983)

On this mission we deployed a satellite for India that incorporates the combined functions of telephone and data communications, direct broadcast television, and weather observation in one satellite. I understand that the satellite is operating successfully today, as are all six of the satellites deployed by the Shuttle to date. The eighth mission also demonstrated the ability of the Shuttle to deploy and berth very large payloads, shown by using the Payload Flight Test Article, which has a mass of more than over 7000 lb (3.175 Mg).

During the re-entry phase of this mission, Astronaut Donald Peterson noticed flashing in the overhead windows. Strapped down, he managed to take advantage of the opportunity and take excellent photographs of this pulsating wake, by pointing his camera in the aft direction. This phenomenon occurred early in the re-entry phase between the altitudes of 400 000 to 250 000 ft (122 to 76.2 km). At this point, the Orbiter travels at approximately Mach 24 and just begins to encounter the upper reaches of the atmosphere. The resulting high-temperature–low-pressure environment causes the air molecules to ionize and a plasma to build up that emits this pulsating sequence.

The Orbiter's glowing nose cone was revealed during landing by the use of infrared cameras. The glow is caused by the intense heating of approximately 2800 °C experienced by the nose cone during re-entry. STS-8 accomplished the first Shuttle night landing, which proved its capability to land at any time, an important requirement when the Shuttle begins landing in Florida where the weather is often better at night.

Delivery of Discovery

The third Shuttle Orbiter, Discovery, was delivered to the fleet on 1 November 1983; it will enable the flight rate to be increased to 10 missions in 1984 and 12 in 1985. A fourth Orbiter, Atlantis, is under construction and will be delivered next December to further expand the Shuttle fleet to meet the ever growing demand for the Shuttle's unique capabilities.

THE SHUTTLE AS A SPACE TRANSPORTATION SYSTEM AND MORE

The STS was initially called a transportation system for a number of different reasons. That label is somewhat unfortunate because it is so much more than that. In addition to providing access to and from space for men and equipment, with its relatively unlimited payload capability through multiple missions, this spacecraft has unmatched versatility of operations once in orbit. Some of these capabilities will now be reviewed.

FIGURE 12. Spacelab 1 configuration.

Spacelab: a laboratory in space

The inauguration of the Shuttle–Spacelab era is the culmination of ten years of effort by the European Space Agency. The first Spacelab mission, STS-9, has just shown how versatile the system is (figure 12). Scientists from Europe, Japan, Canada, and the United States provided over 70 experiments covering astronomy, solar physics, space plasma physics, atmospheric physics, life sciences, and material sciences.

Spacelab was developed on a modular basis and can be varied to meet specific mission requirements. Its two principal components are the cylindrical module, which is pressurized to provide an environment in which astronauts can work without special clothing, and the U-shaped pallets in the back that directly expose telescopes, antennas, and sensors to space.

FIGURE 13. European Space Agency's Spacelab in the cargo hold of the Shuttle.

FIGURE 14. An astronaut wearing an m.m.u. to be used on the Solar Maximum mission.

The Shuttle–Spacelab combination will go a long way towards demonstrating the feasibility of international cooperation in space and will provide the foundation for manned and unmanned space stations of the future (figure 13).

Solar maximum repair

A key Space Shuttle service in orbit is the forthcoming rescue and repair of the Solar Maximum satellite planned for the STS-13 mission in 1984 (figure 14). Activities outside space vehicles by means of a manned manoeuvring unit (m.m.u.) will enable a crew member to fly to and stabilize the Solar Maximum satellite for attachment to the remote manipulator system and placement in the cargo bay. Once the satellite is secured on the payload bay mount, repair work will be initiated and checked. The Solar Maximum satellite will be redeployed by the remote manipulator system and returned to full operation as a solar observatory.

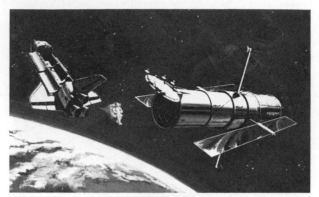

FIGURE 15. To illustrate the servicing capability of the Shuttle.

Space Telescope

Spacecraft such as the Space Telescope will be routinely serviced in orbit for extended use (figure 15). Orbiting spacecraft may have sensors, experimental packages, high resolution film, and other elements retrieved, and may be resupplied with new experiments, film, or provisions.

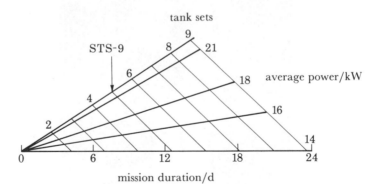

FIGURE 16. Increases in time spent in orbit.

Extended duration Orbiter

The time spent in orbit will be extended as orbital activities evolve; Shuttle users will definitely need to stay in space longer as experimentation becomes more rewarding and complex (figure 16). A cryogenic wafer can be added to the spacecraft that will extend the present 7 to 10 d capability up to approximately 15–18 d. A solar sail and battery combination with gaseous cryogens could provide 30 to 45 d in orbit. This aspect of Shuttle's capability, when added to the Skylab foundation, will assist in parallel the concept and development of permanent manned stations of the future. It will serve as a bridge between now and the date when a permanent station is a reality.

Expanding Shuttle operations

The Shuttle has already proven its ability to deliver and retrieve satellites (figures 17 *a* and *b*). Both of these capabilities will dominate the early operational years of the Shuttle.

Satellite repair and servicing will begin early next year with the Solar Maximum repair mission (figure 17 *c*). This first step will pay increased dividends in the future as satellite designers begin optimizing their satellites not only for repair, but also for routine improvements. The Spacelab era is here, and it will enhance our ability to explore all disciplines on a greatly expanded scale (figure 17 *d*). Such work will lead to the commercial use of the space environment.

Until the advent of the Space Shuttle, the use of space had been limited by launch system capabilities. The Shuttle system has greatly expanded these limits (figure 17 *a–f*). But in addition, the Shuttle provides the opportunity to have human wisdom and will in space. This allows us to perform assembly in orbit and thereby remove size and weight constraints for space structures.

It is my hope that through these advances, we will witness in the coming decades the realization of large-scale international space operations to further our scientific knowledge, to open commercial opportunities, and to enhance and improve life on Earth.

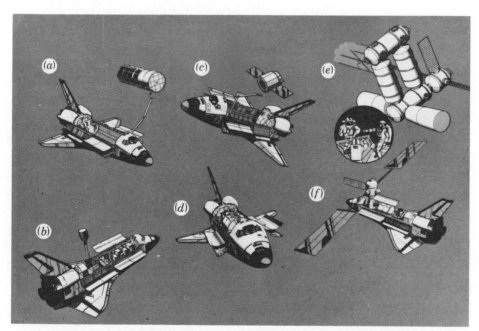

FIGURE 17. The Shuttle's uses: (*a*) retrieval; (*b*) satellite placement; (*c*) repair and servicing; (*d*) laboratory; (*e* and *f*) assembly.

Discussion

F. MILES (*Independent Television News, London, U.K.*). We have recently read and heard many inaccurate reports of a so-called 'emergency' facing Soviet cosmonauts aboard Salyut 7. I hasten to add that although we at I.T.N. are constantly ahead of others in reporting Russian space activities, we never once suggested the cosmonauts' lives were at stake. But we were puzzled by an extraordinary rumour published by others that Columbia was being made ready to rescue the 'stranded' cosmonauts. Could Mr Meechan comment on this and on the feasibility that at some time in the future a Shuttle might be used for a rescue mission and, if it could, how people would be transferred from one craft to another?

C. J. MEECHAN. At this time, the Shuttle Orbiter has no docking provisions to attach to other space structures. In the Apollo–Soyuz programme, we did develop such a mechanism to connect the two spacecraft. This was an androgenous connector, meaning that neither vehicle had a male or female connector (a story in itself). So, it can be done. However, at this point, retrieval of astronauts in space from another vehicle would require that the astronauts had space suits. Then they could be transferred into the Orbiter and returned to Earth.

C. LEDSOME (*The Design Council, London, U.K.*). About fifteen years ago, as one of the final parts of the Apollo applications programme, the company I then worked for, now known as Teledyne Brown, produced a design for a spacecraft docking port that could be docked with itself rather than needing a mating port. This was used on the Apollo–Soyuz mission and the intention then was to fit such a port on all future spacecraft to facilitate rescue from any craft by any other craft. I understand from your previous remarks that such ports are not fitted on the Shuttle. Could you comment please?

C. J. MEECHAN. That is correct, we had such ports for the Apollo–Soyuz mission. We do not have such connectors on the Shuttle, but we have proved that it can be done and we could return to a similar concept in the future.

D. O. FRASER (*British Aerospace p.l.c., Dynamics Group, Space and Communications Division, Bristol, U.K.*). Is there anything being done to develop a recoverable first stage for the Shuttle?

C. J. MEECHAN. There have been studies conducted to recover or further utilize the external tank, which I believe you refer to. It is not practical to recover it, but it may be practical to continue its boost into low Earth orbit and use the structure for other purposes. Also, there always remains some fuel in the tank and this residue might be utilized in orbit. Work along these latter directions continue and I would expect that further utilization of the external tank will occur in this decade.

D. WHITEHOUSE (*Mullard Space Science Laboratory, Dorking, Surrey, U.K.*). Could you comment on the total number of Space Shuttle launches, their frequency and the lifetime of the shuttle system?

C. J. MEECHAN. First, let me say that we have just delivered the third Orbiter, Discovery, to N.A.S.A. and will deliver the final one on order, Atlantis, in late 1984. In 1984, the three Orbiters will be used in 10 launches. In subsequent years we expect the launch rate to grow to over 20 launches per year with this fleet. We plan for 100 mission design life with refurbishment and standard maintenance, which, if all goes well, should take us well into the next century with the current fleet. However, prudent planning would suggest that we should build additional Orbiters in this time period.

Phil. Trans. R. Soc. Lond. B **312**, 103–108 (1984)   [103]
Printed in Great Britain

ERS-1: its payload and potential

By Sir Peter Anson

Marconi Space and Defence Systems Limited, Browns Lane,
The Airport, Portsmouth, Hampshire PO3 5PH, U.K.

European Remote Sensing Satellite Number 1 (ERS-1) is a truly international project promoted by the European Space Agency (E.S.A.) and involving 10 countries. The decisions about what should be the aims of the satellite have been reached as the result of lengthy discussions between E.S.A. and potential customers, and of course have also had to take into account the budgetary limitations imposed by the participating countries. The major objectives of ERS-1 are set out in the paper together with a very brief outline of the capabilities of the instruments required to meet these objectives. The paper concludes by suggesting areas in which there is a need for technological advance, which combined with concentrated marketing activity will ensure a commercial future for the remote sensing capability that will be demonstrated by ERS-1.

1. The aims of ERS-1

The purpose of the ERS-1 spacecraft and its supporting ground infrastructure is to make measurements from which specific products can be made available at sufficient speed to be useful. These products have been carefully established through consultation between expert groups of customers and the E.S.A. They form the requirement for the system and in its turn the ERS-1 payload. The products are now listed.

(i) A map of the surface wind over the complete globe at a spatial interval of 25 km. This product is very perishable and the results must be available to users within hours.

(ii) A global picture of the wave structure sampled at 100 km intervals.

(iii) An image of a large area of ocean or land. This is required to a spatial resolution of 30 m but is only produced when the spacecraft is in view of a suitable ground station.

(iv) Information on sea and ice surface structure by obtaining wave height and the sea surface profile to an accuracy of 10 cm. This operates over the whole globe, and provides a sub-product in the form of valuable data on the ocean currents.

(v) Sea surface temperatures to a high accuracy, available even when there is significant cloud cover.

To summarize, the major products are: global coverage of surface winds; global coverage of sea wave spectra; limited coverage of ocean and land surface imaging; global sea and ice surface structure; global sea surface temperature.

2. Constraints

To collect the necessary data to produce the products, we require full Earth coverage on a regular basis. This can only be achieved by choosing a near polar orbit, and since it is important for many studies to take samples at a constant angle of the sun, or in simpler terms at the same time of day, the orbit is slightly tilted to achieve this object. The orbit slowly precesses so that over a period of three days it is possible with some of the sensors to view the whole Earth. Fifteen and one third orbits are made each day and the spacecraft will fly at a mean altitude of 770 km.

Additionally, it has been decided that the payload must be mated to the platform used by the French satellite 'Systeme Probatoire Observation Terrestriale' (Spot). It is from these decisions that the payload mass, power and data rates and storage statistics can be deduced; they are given in table 1. These would perhaps be meaningless if the comment were not added that the satisfaction of the agreed products within these constraints proves to be an exceptionally demanding requirement.

TABLE 1. PAYLOAD STATISTICS

payload mass	900 kg
payload power consumption	1800 W
X-band data rate in image mode	102 Mbit s^{-1}
X-band data rate in non-image mode	15 Mbit s^{-1}
internal data storage capability	6.5 Gbit

3. THE INSTRUMENTS

The ERS-1 payload has been fitted with measuring instruments (see Haskell 1983) to collect the data necessary to supply the products described in §1.

Active microwave instrument

This is a most complex radar operating at 5.36 GHz in a number of modes to measure the sea surface radar reflectivity, to provide in turn:

the wind force and direction over a 500 km swath;

the wave structure (sea state) every 100 km of track;

a radar image over an 80 km swath and 2000 km length for each orbit.

The last measurement is achieved by using Synthetic Aperture Radar (SAR) techniques and is limited by spacecraft power and data storage capacity so that images can only be made for periods of a few minutes and when in direct contact with a ground station.

Radar altimeter

This instrument measures the delay time of a radar pulse from the spacecraft to the sea surface and back. It also examines in detail the shape of the return pulse, which will have been distorted by the depth of the wave structure. Before details of the sea profile can be calculated, it is necessary to add the geoid shape and the spacecraft orbit into the computation. Accurate details of the orbit are obtained by the Precision Range and Rate Experiment (PRARE) and by using the laser reflector in conjunction with ground laser radars.

In practice the radar altimeter can measure the range from spacecraft to the sea with an absolute accuracy of 2 m and a relative accuracy of 10 cm. It can also measure the wave height over a range of 1 to 20 m to an accuracy of 5 cm. It takes measurements at 7 km spacing and in conjunction with the ground terminals can make its product available in 3 h. This instrument also has a capability of measuring wind speed.

Along Track Scanning Radiometer (ATSR)

This instrument operates in the infrared region of the spectrum and measures the magnitude of incoming radiation from the sea surface. Over a 500 km swath it measures the surface temperature of the sea to an accuracy of 0.5 K and the water vapour content of the atmosphere can subsequently be computed to an accuracy of 3%.

Microwave sounder

This measures incoming radiation, in two bands at 24 and 36 GHz, from the sea surface. The radiation measured by both this instrument and the ATSR is modulated by the small-scale atmospheric properties in their respective spectral bands and this information is used to correct the radar altimeter pulse flight time to that relative to a vacuum condition, to calculate the actual distance between the spacecraft and the sea.

4. DATA TRANSMISSION

The instruments listed in §3 generate a great deal of data that has to be transmitted to the ground for subsequent processing into a useful product. This is achieved with the integrated data handling subsystem, a block diagram of which is shown in figure 1.

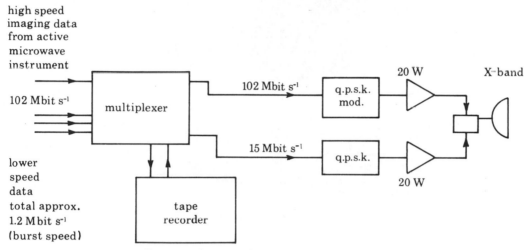

FIGURE 1. Integrated data-handling subsystem.

It can be readily seen that the very high-speed data generated by the imaging radar at 102 Mbit must be passed directly through the subsystem and transmitted to the ground. The lower speed data can be stored on the tape recorder until the next opportunity of contacting a ground station when the data can be released at 15 Mbit s^{-1} and transmitted to the ground.

Figure 2 shows the relation of the ERS-1 payload to the Spot platform and the configuration of the antennas associated with the measuring instruments described. The ground coverage of the sensors shown in figure 2 is illustrated in figure 3.

5. THE POTENTIAL OF ERS-1

It is appropriate to consider the potential of ERS-1 in the light of the title of this symposium. It must be readily apparent that ERS-1 itself will, as soon as it is commissioned, add greatly to the useful data available for weather forecasting, ice reporting, mapping, oceanography and climatology. The results provided by imaging radar are spectacular and can be used to plot the passage of ships as well as oil slicks, and the movements of shallow sand banks. The imaging radar in ERS-1 which is essentially an ocean-observing satellite, could, with small modifications, be used for the purposes of land observation. For this, the potential products could be:

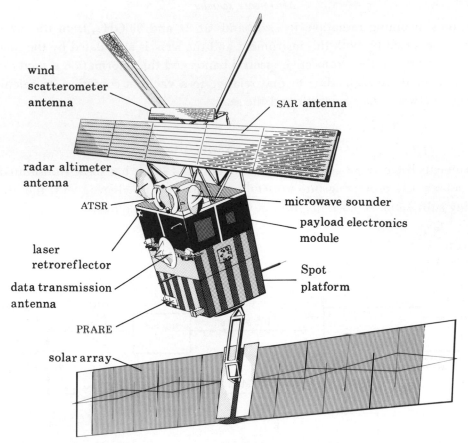

FIGUPE 2. ERS–1.

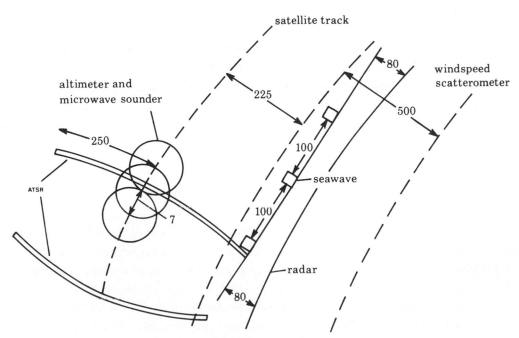

FIGURE 3. ERS–1 coverage. All distances are in kilometres.

the forecasting of crop development; geological surveys in search of oils and minerals; cartography; the tracking of pests such as locusts; and the control of forest diseases such as the recent elm disease. This list is not meant to be exhaustive, but generally indicates the value of the information and the types of customer that should be interested.

6. THE INDUSTRIALIZATION OF ERS-1

If ERS-1 is to be successfully industrialized, it means that its products or other similar required products must be produced so economically that it will be possible to find customers ready to pay a price that will prove profitable to the satellite manufacturer. It is necessary then to examine ERS-1 to establish its weak points from an economic viewpoint and thus to establish the technology targets for the 1990s that will enable successors to ERS-1 to be industrialized.

Swath width

The swath width of the imaging radar is crucial. It is very conceivable that in the years to come, in the interests of maritime law and order, it will be considered desirable to maintain an operational plot of all ships at sea. It can be simply calculated that with a swath width of 100 km and a requirement to image the entire ocean surface four times daily, a system of 120 satellites would be required. This is clearly ridiculous, but if the swath width could be increased to say 600 km, the system would be reduced to 20 satellites and possibly become manageable. If a satellite with a two-sided radar was used the number of satellites could be reduced to 10. A major objective must then be to increase the area covered by the imaging radar of each satellite.

Power limitations on the radar

In ERS-1 the imaging radar is limited to a maximum of ten minutes operation in every orbit of approximately one hundred minutes; this is due to a lack of adequate energy. It should be an objective to ensure that the radar can be operated for at least half of each orbit, although it is probably reasonable to accept that it will not be possible to achieve radar operation when the satellite is in eclipse.

Data transmission

The imaging radar of ERS-1 can only be operated when the satellite is in direct view of a ground station. Ground stations are costly and their number must be kept to a minimum. An objective must be to increase vastly the data storage capability of the satellite, or alternatively to arrange for the processing of the data on board, so that only useful products need to be stored before transmission to the ground. Another solution that should be investigated, and may well be more practical, is to employ a geostationary data relay satellite.

Ground processing

At present the U.K. National Remote Sensing Centre at R.A.E. Farnborough requires more than a day to process a sea area of 25 km × 25 km. It is good to know that new computers have recently been approved, which will greatly reduce that time. However, it is essential that, if remote sensing is to become commercial, the processing of the data, either on board the spacecraft or on the ground, will need to be achieved in real time or at a speed of this order.

Instrument coordination

It would appear a very important technological challenge to coordinate the product of a number of instruments to achieve a simple rapid and reliable conclusion. One could envisage that to achieve a successful prosecution of a ship illegally pumping bilges and causing pollution would require the combined product of an imaging radar, an optical ocean colour monitor and possibly an additional instrument to examine the infrared signature of the ship and hence narrow down its identification. As such matters will be legal in nature the accurate correlation of the products of all the imaging instruments will have to be proved beyond doubt.

Marketing

While the necessary technological steps are being taken to make the ERS-1 concepts economically viable, there must be a major corresponding effort to educate potential customers in the facilities that will be available towards the end of the decade.

7. CONCLUSION

The timescale for ERS-1 is: beginning of 1984 to end of 1987 for design and manufacture; launch at the end of this period; exploitation from the time of launching to the end of 1990. It appears essential now to start an ERS-1 industrialization preparatory programme, aimed at tackling the problems of swath width, power limitation, data processing and storage and instrument coordination, previously mentioned. At the same time it is the duty of all of us who have been privileged to learn early about the potential of remote sensing to spread the gospel by all means, such as lectures and articles so that the exploitation phase of ERS-1 from 1988 to 1990 will prove to be the culmination of a successful marketing campaign that will lead to the commercial exploitation of improved ERS-1 facilities in the early 1990s or possibly sooner.

Since this paper is a review it contains much material from unpublished papers produced by my colleagues working for the companies and organizations engaged in the European Remote Sensing Programmes. I am indebted to them for this material.

REFERENCE

Haskell, A. 1983 The ERS-1 Programme of the European Space Agency. *ESA Jl* **7**, 1–13.

Discussion

D. G. STEPHENSON (*Commercial Space Technologies Ltd, Hanwell, London, U.K.*). Will the radar altimeter on ERS-1 be able to detect the ground level swellings that occur over sub-surface magma accumulations and which can be the precursors of volcanic eruptions? A current example being the resort of Mammoth Lake, California. If so, has the possibility of selling these data to local planning and disaster coordinating agencies been examined?

SIR PETER ANSON. The radar altimeter on ERS-1 has been designed to work only over the sea. It will not therefore be able to detect the ground level swellings, which can be the precursors of volcanic eruptions.

Phil. Trans. R. Soc. Lond. A **312**, 109–114 (1984)   [109]
Printed in Great Britain

The processing and use of data from Earth observation satellites

By D. D. Hardy

Remote Sensing Division, Royal Aircraft Establishment, Farnborough, Hampshire GU14 6TD, U.K.

Most systems reliant on advanced technology present a familiar dilemma: the system designer does not know what the customer wants, while the customer does not understand the technology well enough to know what is possible. Although Earth observation satellite systems ought ideally to be designed for all customer needs, this is impossible for several reasons. Not least of these is the difficulty of identifying at the outset all, or even most, of the possible customers.

This circumstance makes the creation of Earth observation systems somewhat speculative and imposes particular constraints on the subsystems for processing and use of the data. This paper discusses the technical and institutional aspects of processing and dissemination of data from remote-sensing satellites for the benefit of the user.

Introduction

The first governmental step towards the industrialization of Earth observation by satellite was taken in 1977, when the Department of Trade and Industry financed the setting up of a centre of excellence in space remote-sensing technology at the Royal Aircraft Establishment. The National Remote Sensing Centre, which is closely associated with the Department of Trade and Industry programme, has been in operation at Farnborough since 1980. In this paper the theme of the meeting is discussed in the light of experience gained from participation in this work.

At the beginning of 1977 there was some research into applications of remotely sensed data going on in universities and elsewhere; a few firms were already applying the data and offering added value services. It was therefore decided that the main thrust of the R.A.E. work should be towards acquiring, processing and disseminating Earth observation data from spacecraft and extension of the technique towards new applications and new users.

It is a truism that the introduction of advanced technology into practical applications is fraught with difficulty. Although reluctance to accept novel systems is usually (and often correctly) ascribed to conservatism, there is in fact a very real dilemma. The system designer is likely to be motivated by 'technology push' and the desire to produce a technically elegant solution. The dilemma arises from the difficulty of defining the problem. The customers may not know what is technologically possible; indeed, most do not know that they are potential customers. Lack of communication between designer and user has in many fields led to the creation of a number of clever and expensive white elephants.

In Earth observation from space this dilemma is particularly acute. The technology is both advanced and expensive: gestation of a new system is protracted; and the range of potential users is large in number, in diversity and in the extent of their technical understanding. Although the design of Earth observation systems is usually preceded by attempts to identify, and to define, most of the possible applications, past remote-sensing satellite missions have invariably led to a number of unforeseen applications. Clearly, there must be many potential applications of existing systems that have yet to be identified, and of course the advances in sensor and supporting technologies will create still more.

This circumstance makes the creation of Earth observation systems somewhat speculative and imposes particular constraints on the subsystems concerned with the processing and use of the data. These constraints affect both the technology to be employed and the institutional structure for acquisition, processing and dissemination.

DATA ACQUISITION

The high investment cost of space-based Earth observation is primarily justified by global coverage and hence the potential recovery of those costs from a very large number of users. The inevitable corollary of this is that enormous quantities of data are generated; the output of existing imaging sensors is of the order of 10^8 bit s^{-1} or, if operated continuously, almost 10^{13} bit d^{-1}. Of course, the volume can be reduced by the adoption of sampling strategies and the use of so-called smart sensors or systems that reject unwanted data. In the present ill defined state of the market, the tendency has been to capture all the data that might eventually be useful, and I predict that techniques that irreversibly restrict geographical or temporal coverage will remain unpopular for some time to come. The implications of this, together with improvements in sensor performance, are that data throughput will increase by an order of magnitude by the early 1990s.

The large size of the data sets, and the rate at which they are generated, tend to determine the design of the ground segments of Earth observation systems. High bit-rates occupy large bandwidths in data transmission from spacecraft to ground and from ground station to user. Data processing often involves the manipulation of large matrices, and the configuration of the facilities used is usually dominated by input, output and storage requirements. The requirement for archives, both in data storage technology and in system architecture, is extremely demanding. The other major influence on the ground segment is the timescale in which the processed data are required; the nearer to real time the requirement and the larger the area to be covered, the more expensive the operations are likely to be. If these constraints are related to the existing state of the art the following scene emerges.

Data recorders on-board unmanned satellites have capacity for only a very small part of the potential coverage per orbit of high resolution imaging sensors. The output of such instruments must either be transmitted to a network of ground receiving stations or must be relayed by satellites in higher orbit to one or two ground stations. The N.A.S.A. Tracking and Data Relay Satellite System will use two satellites in geostationary orbit to transmit to a single ground station in the U.S.A. to achieve almost global coverage. Although such a relay system is expensive, so is the alternative of a network of direct receiving stations. Because of the large signal bandwidth it is necessary for transmission systems to operate at microwave frequencies. Typically, ground station antenna apertures of 10 to 12 m are needed, and the resulting narrow beamwidth combined with the tracking requirements for low Earth orbit call for high angular rates of antenna motion and sophisticated control systems. Reception, demodulation and demultiplexing pose no particular problems, but data capture (i.e. recording) and real-time processing may well do so. Serial multitrack magnetic tape recorders are available, which can accept data at rates approaching 10^9 bit s^{-1}, but investment and running costs are high. Further, serial records cannot be accessed directly by conventional computers. The use of helical recording techniques may permit the development of computer addressable serial recorders for high bit rates, but progress has been slow.

DATA PROCESSING

At present, applications of Earth observation that require real-time or near real-time delivery of data (such as meteorology and oceanography) can be satisfied by data from sounders and from imaging sensors of modest spatial resolution. The data handling and processing functions for these can be performed in real time by conventional mini- or micro-computers. However, as the AGRISIPINE experiment made in 1982 demonstrated, there is a range of potential time-dependent applications that can only be satisfied by supplying high resolution processed data within a few hours of acquisition. All-weather day and night sensing by synthetic aperture radars (SARs) will add a further quantum of data processing capacity to the ground segment specification. The current approach to processing of large data sets from imaging sensors is to use one or more of the faster minicomputers, in conjunction with special purpose pipeline or parallel hardware processors for particular operations. These are unable to cope with the throughput in real time and the standard technique is to replay the original serial record at a fraction of the record speed to slow down the operation. Typically, the record:replay ratio is limited to 16:1 or thereabouts; this is insufficient reduction for most SAR processors currently available and a further stage of transcription is necessary for this sensor.

The advent of powerful processors 'on a chip' and equally compact random access memories will permit the development of faster data processing systems in which the architecture is optimized for specialized applications, but which can employ standard computer peripherals and communications interfaces. It will be necessary to ensure that fast processors are not, as at present, constrained by input–output limitations.

The processes performed at the ground station are what is usually termed pre-processing; this consists of the removal of artifacts of the sensing process, corrections by using pre-flight or in-flight calibration data, and location of the data set on the Earth's surface by using prior knowledge of the satellite orbit.

A further phase, usually termed precision processing, consists of radiometric and geometric correction by using *post hoc* information such as refined measurements of the satellite orbit and attitude and ground control points. This work is usually done in specialized centres such as the National Remote Sensing Centre. The output is normally in image form, transformed into the coordinates of a specific map projection, and may be supplied to the user as a computer-compatible digital record or a photographic image.

At present, archives are formed of the data at each of these stages; a primary archive on serial tape at the ground stations and secondary archives of pre- and precision-processed data at National Centres. It may be foreseen that operational remote-sensing systems will adopt a rather different approach. The needs of time-dependent users will be met by on-line processing on-board the satellite or at the ground stations, and precision-processed products will be generated as standard for all other users. All archives will be held in computer-addressable form and the community using remote sensing will continue, with others, to urge the computer industry toward the adoption of cheap high-density recording media such as 'laser discs'. Further development will be needed to produce a better recording system for data capture at the ground stations.

For experimental missions, and for archiving of meteorological data for research purposes, sampling strategies will have to be evolved to reduce the volume of data to manageable proportions.

ANALYSIS AND INTERPRETATION

The ultimate customer of operational remote sensing is unlikely to be able to make direct use of an image or tabulated data, and may be satisfied by something as simple as a 'yes or no' answer, a date or a few numbers. The most frequently used technique for transforming images into such information is still visual interpretation. I predict that this method will continue to be used for many years, but that increasingly sophisticated machinery will be deployed to aid the interpreter. In parallel, knowledge-based systems will increasingly be used for operational analysis and interpretation and ultimately the two strands of development will tend to merge. One can envisage an expert system initially 'learning' from a skilled human interpreter and subsequently developing faster and more efficient procedures. In the meantime there is a substantial market for interactive analysis hardware and software ranging from the fast, flexible and expensive to the slow, simple and cheap. The provision of facilities for remote sensing to be used in secondary and even primary education is an essential prerequisite of its industrialization.

There is, of course, a market for those skilled in analysis and interpretation to sell their services to those who are not. It may be, however, that, as for telegraphists and telephone operators, the advance of information technology may eventually reduce the number of jobs in this area.

DISSEMINATION

In pursuing the flow of data processing from ground station to user, the means of data transmission has so far been ignored. Although, during pre-processing, there is a natural tendency for the volume of unique data to be reduced (for example, cloud obscured or poor quality data are discarded), the replication of data sets for a number of users tends to maintain or increase the total volume to be disseminated.

At present, most dissemination is by telephone lines for meteorological data, and by post for computer-compatible tapes of data of high resolution. Both of these methods have their limitations. The SPINE experiment demonstrated that the routine delivery of Landsat data to national centres in the U.K. and Sweden by a 2 Mbit s^{-1} satellite communications link was beneficial to many users in those countries. The AGRISPINE fast delivery experiment was also successful in supporting several time-dependent applications during the agricultural growing season in 1982. This transmission rate has been adopted for the terrestrial Metsatnet communications network, which will be installed in 1984 to transfer environmental data within the U.K. It seems likely that, where time is of the essence, international distribution by satellite, and internal dissemination by wide-band terrestrial networks, will become the norm. The services offered by common carriers will readily accommodate these requirements.

For other users, postal distribution of computer-compatible tapes, floppy and video discs are likely to persist, but increasing use will be made of controlled broadcast via cable networks and of dial-up facilities for small users to gain access to publicly available archives. Such archives will have to be carefully designed to meet the needs of a wide range of future users.

This broad and necessarily superficial review might be summarized by stating that the technical problems likely to arise in the industrial exploitation of remote sensing from space are not trivial, but solutions to all of them can be foreseen within current technological developments. That is not, of course, to say that the solutions will necessarily follow precisely the forecasts given in this paper.

TABLE 1. ORDER OF COST OF ENTRY

	$£$
space segment operator	over 10^8
ground segment operator	over 10^7
digital data analyst and interpreter	10^4-10^6
visual interpreter	under 10^2
teacher	*ca.* 10

INSTITUTIONAL ASPECTS

The institutional problems will be harder to solve. Their scope is illustrated in financial terms in table 1, which indicates the order of cost of entry into the remote sensing business for a range of possible participants. For particular applications, where existing sensors can already offer quantifiable benefits, national, public or private sector ownership of space segments is possible. However, for experimental missions and for operational missions employing advanced sensors, such as multi-spectral radars, it seems likely that international initiatives will be needed.

What is quite certain is that encryption of transmitted data will become standard practice, and that the owners of space segments will thereby exercise some control over ground segments and over the exploitation of the data. This could include manipulation of the market for ground station equipment, withholding commercially valuable data or simply maximizing revenue. It is thus necessary for the U.K. to participate effectively in both space and ground segments if continued availability of data to U.K. industry is to be assured. For the space segment, membership of the appropriate international clubs may perhaps suffice, but in any case a powerful national capability for acquiring, processing, archiving and disseminating data is clearly essential. It should be noted that 'disseminating' to end users is synonymous with 'selling'.

The extent to which these facilities should be publicly or privately owned is outside the scope of this paper, but there are a number of factors to be considered in this context. The first is the relation between the market and the technology, mentioned in my introduction. The rest arise mainly because the investment needed for a comprehensive ground segment is substantial, not only for hardware but also for the research necessary before novel sensors can be fully exploited. On the other hand, many of the potential customers will make only a small contribution to investment and operating costs, while requiring a disproportionate amount of technical advice and support. A national facility has to provide the support if the use of the technique is to grow. The establishment and maintenance of a national global archive also requires substantial investment while generating rather small revenue, at least in the early stages.

There are currently just over 250 active customers of the U.K. National Centre and they are divided almost equally between government, industry and academe. All groups make use of data from the U.K. and overseas, and it would appear that most of the major revenue-earning applications in the industrial sector are overseas. Although these proportions have persisted throughout the existence of the Centre, there are now some indications that, while each group is growing in absolute terms, the industrial customers are becoming relatively more numerous. The Centre, in supporting industrial initiatives, has to maintain strict impartiality and confidentiality when, as often happens, it is concurrently supporting several competitors for the same contract.

The prices currently charged for data and data products, which ultimately are determined by the space segment operator, are non-optimum. They do not provide a realistic return to the operator; they are high enough to deter the academic researcher or teacher, while being so low in the perception of the industrialist that he is inclined to doubt the potential value of the product. This situation will change rapidly as satellite owners seek to recover more of their costs, and the academic community is likely to suffer in consequence. Clearly, steps must be taken to permit research and teaching to continue in this new cost environment. It has been suggested more than once that payment for commercial use of data should be by way of levy on profits achieved from exploitation, but it is difficult to see how this could effectively be regulated on a global scale.

Another issue to be faced is the distribution of ground facilities; the provision of data acquisition stations is likely to be centralized while analysis and interpretation facilities tend to proliferate. This is quite natural but, as has been described above, there are a number of functions interposed between these interface points for which the decision is less clear cut. Moreover, technological development is likely to change the constraints that have given rise to contemporary arrangements.

Whatever the ultimate national infrastructure, and whatever its ownership, it must not inhibit the individual or corporate entrepreneur from exploiting the technique. Further, it must seek to create opportunities for such local initiatives while pursuing international partnerships.

Conclusion

The objective of this paper is to promote discussion. From a particular viewpoint, a number of technical and institutional problems have been identified. Other relevant issues will be perceived by others with different perspectives. The successful industrialization of Earth observation from space will depend on sensitive resolution of these issues and of the often conflicting objectives of their proponents.

Discussion

D. O. FRASER (*British Aerospace p.l.c., Dynamics Group, Space and Communications Division, Bristol, U.K.*). In view of the problems of world coverage for SAR imaging referred to by Sir Peter Anson and the need for coordination between remote-sensing satellites, does Mr Hardy think an international-remote sensing organization, comparable to Intelsat and Inmarsat, may be formed?

D. D. HARDY. That is certainly a possibility, but there is a countervailing tendency (Spot, Landsat) toward national and private sector investment in remote sensing, which may, or may fail to, lead to operational systems. There is also a tendency for the more ambitious and complex pre-operational systems – of which Radarsat is an example – to be organized on a bi- or trilateral basis.

Coordination does not necessarily imply multilateral operating entities and I would point to the work on data format standardization that has already been done by informal international groups.

In the end, one has to balance the inherent inefficiency of multilateral bodies against their political and financial advantages. At present there is no apparent will to create such a body for remote sensing.

Phil. Trans. R. Soc. Lond. A **312**, 115–118 (1984)   [115]
Printed in Great Britain

Global Habitability and Earth remote sensing

By S. G. Tilford

*Earth Science and Applications Division, National Aeronautics and Space Administration,
Washington, D.C., 20546 U.S.A.*

[Plate 1]

This paper discusses current techniques developed by N.A.S.A. for observation of the atmosphere, oceans and land from satellites. N.A.S.A. has developed a new concept called 'Global Habitability' to provide a framework for an interdisciplinary, global scientific research programme. Its aims are discussed together with future projects for Earth remote sensing.

In 1960 N.A.S.A. launched the Tiros satellite into polar orbit to study the atmosphere of the Earth. Later in the 1960s, hand-held cameras on the first manned flights gave us additional new and exciting glimpses of the Earth. In the early 1970s N.A.S.A. flew sensors optimized for observing the oceans on the Skylab mission, and the launch of the Earth Resources Technology Satellite (ERTS-1) in 1972 gave us the means of studying the land masses. So, by the early 1970s we had demonstrated the ability to observe the atmosphere, oceans and land areas over practically the entire Earth.

Data acquired from space are now being used on a regular basis for research in agriculture, land use, hydrology and geology, and the value of these data has been fully demonstrated. We can now make land cover inventories, locate, classify and measure major forest types, identify shoreline changes, salinity zones and flood plain boundaries, and identify water impoundments greater than two acres (8084 m²). The knowledge gained from these studies has been used to forecast commodity production for wheat and to identify crop stress areas. These capabilities can be used to assist in determining global biomass changes and in conjunction with a global biology programme can begin to determine the process by which changes in biomass occur.

Perhaps the most dramatic example of the use of space-based data has been for atmospheric physics. Research instruments and techniques developed by N.A.S.A. to obtain global measurements of tropospheric parameters such as temperature, humidity and winds have led to a substantial improvement in our knowledge of the troposphere. The same techniques have been incorporated into National Weather Service models to improve the quality of long-term forecasts. Data from recent N.A.S.A. satellites are being used to update forecasting techniques so that we may approach the goal of a reliable seven day forecast. N.A.S.A. satellites have also measured the distribution of the ozone and other important species in the stratosphere. These data have improved our basic understanding of the stratosphere and of the role that man-made pollutants might have on the ozone content. An experiment carried on the Shuttle has demonstrated a capability to measure a tropospheric trace species (carbon monoxide) from space.

It is now possible to measure both sea surface winds and wave heights, and to determine the circulation patterns in the oceans by using satellite techniques. Measurements of the chlorophyll content of the sea can now be made routinely and have been used to determine the probability of fish harvests off the West coast of the United States. Improved algorithms have been developed to determine global skin surface temperatures (figure 1, plate 1) from the current operational NOAA polar orbiting satellites.

Each step that has been taken forward in specific disciplines in remote sensing has provided another piece of a very large puzzle that is the Earth system. It is now time to assemble the separate pieces into a coherent whole. Over the past year, N.A.S.A. has been working with other organizations to develop a concept called 'global habitability', to provide a framework for an interdisciplinary, global, scientific research programme.

The objective of the programme is to investigate long-term physical, chemical, and biological trends and changes in the Earth's environment, including its atmosphere, land masses, and oceans. The programme will specifically investigate the effects of natural and human activities on the Earth's environment by measuring and modelling important physical, chemical, and biological processes and their interactions, and will estimate the future effects on biological productivity and habitability of the Earth by man, by other species, and the effects on natural causes. The programme will involve the acquisition and analysis of space and sub-orbital observation, land and sea-based measurements, modelling and laboratory research, and supporting data management technologies over a ten year or longer period of time. The programme will, of necessity, involve major international participation and will be coordinated as far as possible with national research programmes and those of international agencies. The aim of the N.A.S.A. global habitability concept is to obtain a solid body of knowledge from which policy decisions can be made and which will address the questions of change, either natural or of human origin, effecting the habitability of the Earth.

A primary challenge of the 1980s is to improve upon the quality and usefulness of remote

---

Description of Plate 1

FIGURE 1, plate 1, is the first global map ever made of Earth's mean skin surface temperature. It was derived from multispectral data measured by satellites. The ocean and land temperature values have been averaged spatially over a grid 2.5° latitude by 3° longitude and correspond to the month of January 1979. The mean temperature values for this month clearly show several cold regions, such as Siberia and northern Canada, during winter in the Northern Hemisphere and a hot Australian continent during summer in the Southern Hemisphere. Mountainous areas are clearly visible in Asia, Africa and South America. The horizontal gradients of surface temperature are displayed on the map in colour contour (density of grey here) at intervals of 2 K and show some of the major features of ocean surface temperature, such as the Gulf Stream, the Kuroshio Current and the local temperature minimum in the eastern tropical Pacific Ocean. The sea surface temperatures derived by satellite are in very good agreement with ship and buoy measurements.

Surface temperature data are important for weather predictions and climate studies. Since the cold polar regions cover a small area of the globe relative to the warm equatorial regions, the mean surface temperature is dominated by its value in the tropics. The resulting mean skin surface temperature during January 1979 is calculated to be: global, 14.14 °C (57.46 °F); N. Hemisphere, 11.94 °C (53.49 °F); S. Hemisphere, 16.35 °C (61.43 °F). Currently, climate scientists are testing the accuracy of using surface temperature anomalies in the Pacific Ocean as potential predictors, on seasonal timescales, of weather conditions over parts of North America. In addition, day and night variations in the surface temperature can be used to study soil moisture.

The image was obtained by a team of N.A.S.A. scientists from the Jet Propulsion Laboratory in Pasadena, California, and the Goddard Space Flight Center in Greenbelt, Maryland. The satellite data were acquired by the high resolution infrared sounder and the microwave sounding unit, both instruments flying on board the National Oceanic and Atmospheric Administration (NOAA) weather satellites. The surface temperature was derived from the 3.7 μm window channels in combination with additional microwave and infrared data from the two sounders. The combined data sets were computer processed, by using a data analysis method that removed the effects of clouds, atmosphere, and reflection of solar radiation.

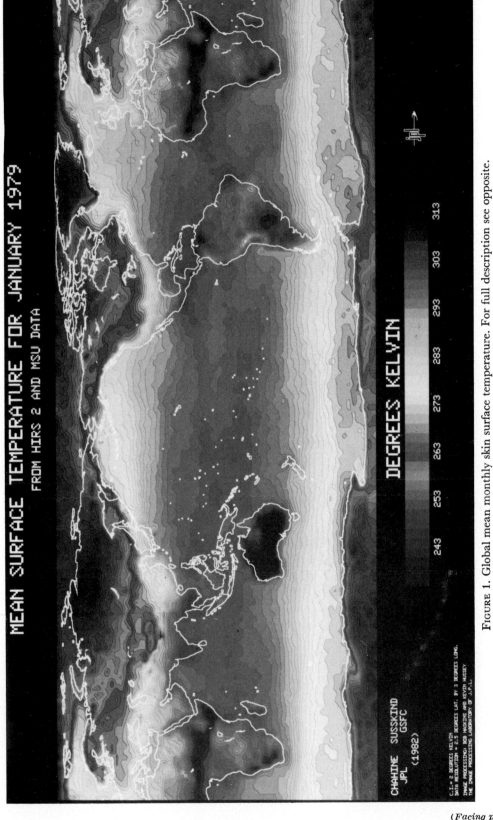

FIGURE 1. Global mean monthly skin surface temperature. For full description see opposite.

sensing information. Landsat-4, launched in July 1982, represents a major step forward in meeting this challenge. The improved spatial, spectral, and radiometric characteristics of the thermatic mapper have already contributed to improved analysis of data from agricultural, wetlands, and geological test sites. The Space Shuttle is also providing exciting new possibilities in Earth observation. The Shuttle Imaging Radar (SIR-A) acquired more than ten million square kilometres of surface imagery during a two day mission in November 1981. An L-band Synthetic Aperture Radar (SAR) obtained imagery of many arid and tropical areas of the world for the first time. The Shuttle also carried SMIRR (Shuttle Multispectral Infrared Radiometer), which has already provided information leading to the discovery of an area of potential metallic mineralization in Mexico. In addition to these recent accomplishments, we are planning more Shuttle-based remote sensing experiments in the 1980s. These include the large format camera in early 1984; the Shuttle imaging radar-B in the summer of 1984; and the multispectral linear array in the late 1980s.

N.A.S.A. is also improving its capability for global studies of the atmosphere and the oceans. The Upper Atmosphere Research Satellite (UARS) will provide scientific understanding of the complex physical and chemical interactions that maintain the ozone layer in the Earth's stratosphere. We are also planning the Ocean Topography Mission (TOPEX) to improve our understanding of the circulation of the ocean. We are developing an advanced radar scatterometer (SCATT) for more accurate sea surface wind measurement, and an Ocean Colour Imager (OCI) for improved primary ocean productivity measurements.

Observations from the Nimbus-6 and Nimbus-7 Earth Radiation Budget (ERB) instruments and operational NOAA satellites are being used as a foundation for developing a continuing series of Earth radiation budget (solar constant) data sets. The data sets formed from these observations will serve as a continuing resource for climate research. These data sets will be continued and augmented by launch of the Earth Radiation Budget Experiment (ERBE) scheduled for 1984. Recent evidence from the Nimbus-7 and SMM satellite observations confirm natural variations in the total solar output of several tenths of 1% for periods of up to about two weeks. To determine the impact of such variations on the climate systems as well as to monitor their long-term trend, several instruments including the Active Cavity Radiometer (ACR) have been designed for Shuttle operations. In addition, instruments such as the Atmospheric Observations from Shuttle (ATMOS) infrared spectrometer have been developed to measure stratospheric and mesospheric trace species. We eventually plan to combine these instruments in a single payload known as the Earth Observations Module.

N.A.S.A. has come a long way in the twenty-three years since the first Tiros. We now have the capability to take the next great step. Building upon the remote sensing advances of the past twenty years, we can now study Earth as an interlocking system of complex global processes. The potential for progress in this next step will be limited only by our imagination and perseverence. We in N.A.S.A. intend to do our part in making the next twenty years in remote sensing as exciting and successful as the past twenty years have been.

Discussion

J. R. PAGE (19 *Kingsland Road, Alton, U.K.*). The limited resolution of currently available visual data from N.A.S.A. satellites for civilian use limits the scientific disciplines that can benefit from remote Earth sensing. Does Dr Tilford think, therefore, that there is a possibility of improving the quality of data in the future?

S. G. TILFORD. I think that the spectral resolution will increase by orders of magnitude with the instrumentation that we now have planned for the 1990s within the Earth Science and Applications Programme, i.e. we expect to fly the Shuttle Imaging Spectrometer Experiment (SISEX) with over a hundred spectral bands at approximately ten nanometre resolution. With respect to spatial resolution, I do not believe that we will develop, in our programmes, footprints with resolutions less than about 15–20 m. With spatial resolutions less than these values, we could have problems with full global distribution and use of these data.

Phil. Trans. R. Soc. Lond. A **312**, 119–131 (1984)   [119]
Printed in Great Britain

Space stations and their potential uses

By R. F. Freitag

Space Station Task Force, N.A.S.A., Washington, D.C. 20546, U.S.A.

This paper discusses the status of N.A.S.A's efforts to define the scope of a space station programme. It presents N.A.S.A's rationale for a space station, architectural, techno- logical and operational concepts that would enhance its capabilities, status of planning, how it would be used, opportunities for commercialization, and the potential for international participation in the programme.

1. Introduction

A process is under way in the United States to decide on a course of action regarding the next major step for the United States Manned Space Programme; that is, should the National Aeronautics and Space Administration (N.A.S.A.) undertake the development and deployment of a permanently manned space station. A positive decision will set the course for space activities for the next two decades not only for the United States, but also for other countries and organizations that could become involved in the programme.

N.A.S.A. has received continued encouragement and support from the U.S. President for a strong, national space programme. The following excerpts of remarks President Reagan made on 19 October 1983 at N.A.S.A's 25th Anniversary Celebration in Washington, D.C., illustrate his strong commitment to space activities.

'Right now we're putting together a national space strategy that will establish our priorities and guide and inspire our efforts in space for the next 25 years and beyond. It will embrace all three sections of our space program – civil, commercial, and national security. The strategy should flow from the national space policy I announced on July 4 last year.

We're not just concerned about the next logical step in space. We're planning an entire road, a 'high road' if you will, that will provide us a vision of limitless hope and oppor- tunity, that will spotlight the incredible potential waiting to be used for the betterment of humankind...'

'...Let us demonstrate to friends and adversaries alike that America's mission in space will be a quest for mankind's highest aspirations: opportunity for individuals, cooperation among nations and peace on Earth.'

This presentation to The Royal Society will allow N.A.S.A. to inform its friends in the United Kingdom its plans for its next major programme: the Space Station.

2. Background

Even before the Apollo programme had successfully landed a man on the moon and returned him safely to Earth, N.A.S.A. engineers and scientists were considering the next step of the space age. That next step was defined to be the simultaneous development of two complementary capabilities. One was a safe, reliable transportation system that could provide routine access to space. The other was an orbital space station, where humans could work and live in space, and which could be the base camp from which other, more advanced 'next steps' could be initiated. This thinking set the stage for two of the most significant activities conducted by N.A.S.A. in the 1970s and 1980s; that is, Skylab and the Space Shuttle.

On 14 May 1973, the United States launched its first space station, Skylab, aboard a Saturn V booster. Skylab demonstrated that humans could function in space for periods up to 12 weeks and, with proper exercise, could return to Earth with no ill effects. In particular, the flight of Skylab proved that man could operate very effectively in the weightless environment and that it was not necessary to provide artificial gravity to live and work in space. Skylab crews accomplished a wide range of emergency repairs on station equipment, which included freeing a stuck solar array panel, replacing rate gyros, and repairing a malfunctioning antenna. The crews twice installed portable sun shields to replace one lost during Skylab's launch. These activities clearly demonstrated the utility of man in space.

Another product of the Skylab programme was N.A.S.A's realization that extravehicular activity (e.v.a.) could be considered normal and routine. Skylab amassed more than 82 h of e.v.a. Skylab also demonstrated the effectiveness of a scientific laboratory orbiting the Earth. Skylab's eight different solar telescopes provided a quantum step in our understanding of the Sun. Materials-processing experiments done aboard Skylab produced unusually high quality crystals of key microelectronic materials. Skylab-generated, Earth resources photography demonstrated the practicality of using space to observe the Earth with sensors tuned to visible, infrared, and microwave regions of the spectrum. Skylab allowed a comprehensive series of life sciences experiments that provided the basic data needed to design equipment and procedures for long-term living and working in space. In short, the Skylab programme was a resounding success and established a baseline of U.S. experience in the operational use of a space station.

Unfortunately, Skylab was not designed for permanent presence in space. It was not intended to be serviced on orbit, although the Skylab crews were able to make certain repairs. It was not equipped to maintain its own orbit, which eventually caused its fiery demise. It was not designed for evolutionary growth and, hence, was subject to rapid technological obsolescence. These shortcomings, as we shall see, are carefully accounted for in the current N.A.S.A. planning for a permanent space station.

While the U.S. was conducting the Skylab programme, the Soviets were embarking on a more ambitious endeavour. The Soviet Union launched its first space station, Salyut 1, in April 1971. In the twelve years since that launch, additional Salyut stations have been placed in orbit. Their programme is an aggressive and expanding national, manned space flight programme combining Soviet military and civilian missions. The two most recent Salyuts, Salyut 6 and 7, represent a second-generation version of the space station. Salyut 6 demonstrated the ability to refuel the station by using the Progress logistics craft. Salyut 6 cosmonaut crews completed missions lasting 96, 140, 175 and 185 days in space. Ten international crews were hosted, representing the nations of Czechoslovakia, Poland, East Germany, Bulgaria, Hungary, Vietnam, Cuba, Mongolia, Romania and France.

Clearly, the Salyut space programme represents a long-term, methodical development of space capabilities that the Soviet Government considers worthwhile and in their national interest. Their programme is obviously well financed and has involved dozens of high energy launches. In the next few years, U.S. space experts expect to see the flight tests of a Soviet Space

Shuttle, the test of an extremely powerful Soviet space booster, and the appearance of a small, highly manoeuvrable space plane capable of re-entry through the Earth's atmosphere. The challenge of the Soviet space programme in general and of the Salyut programme in particular is unmistakable.

In the early 1970s, the United States embarked on the development of a reusable launch vehicle, the Space Shuttle. The Shuttle has completed its orbital flight tests and was declared operational on 4 July 1982. Since then, the Orbiter Columbia and its sistership, the Challenger, have completed four operational flights. The third Orbiter, Discover, has been delivered to the Kennedy Space Center and is being readied for its maiden flight. The fifth operational flight, STS-9, is in space as this paper is being written. The STS-9 flight includes the Spacelab mission that is designed and built by the European Space Agency and marks Europe's first major entry into a manned space programme. The crew on the STS-9 includes a West German scientist, Dr Ulf Merbold. A key element of the Spacelab is the instrument pallet, which was designed and built by the British Aerospace Corporation. The pallet carries experiments and equipment built by other United Kingdom firms.

With the Shuttle Transportation System operational, it is N.A.S.A's belief that the development of a space station should now be initiated as the next logical step for a major N.A.S.A. technological initiative.

3. WHY A SPACE STATION NOW?

A space station is essential to ensure civil leadership in space for the United States during the 1990s. A permanent, U.S. manned space station is needed to maintain the continuity and focus of the U.S. space programme and to present a continuing challenge to industry and government.

TABLE 1. FUNCTIONS OF A SPACE STATION

on-orbit laboratory	communications and data processing node
science and applications	permanent observatory
technology and advanced development	transportation node
servicing facility	manufacturing facility
free flyers	assembly facility
platforms	storage depot

The functions of a space station, which is a multi-purpose facility, are listed in table 1. The basic need for a space station is associated with providing a permanent facility in orbit that is rich in power, has ample astronaut, scientist, and technician man-hours available, and provides large facilities to assemble, deploy, and operate very large spacecraft that are too large to be carried in the Orbiter payload bay.

A prime purpose of a space station will be the establishment of a laboratory in orbit to capitalize on the unique environment of space, one of the properties of which is weightlessness (sometimes referred to as microgravity). The space station will be a base for upper stage and spacecraft deployment, servicing, and retrieval, which allows the concept of reusable upper stages and spacecraft to reduce costs just as the reusable Orbiter has reduced launch costs. In the future, the space station could allow the creation of immense spacecraft assembled in orbit from subassemblies or pieces brought up by Orbiters, extremely large antennas, power generation equipment, large spacecraft for Earth departure expeditions, and large manufacturing plants.

A space station would extend and enhance the effectiveness of our national security space assets. At present, the U.S. and other countries use space as an arena in which essential national security activities are conducted. The United States employs spacecraft for a variety of defence missions including early warning, communications, navigation, and surveillance. Late in the 1990s, it might be appropriate to deploy a space station in polar orbit belonging to the Department of Defense. Such a space station could build upon the lessons learned from the initial civilian space station and could be designed to satisfy national security requirements of that period.

The space station programme will present challenges that will create technological advances in many disciplines, which will lead to benefits in sectors of our society far removed from the space programme. Early studies of the needs and mission requirements of a manned space station in the 1990s indicate that we will require advances in such key areas as space suit capability, data management and processing, electrical power generation, storage and distribution, and thermal control systems design. The same contractors who will develop and supply these technologically advanced capabilities for a space station will soon be using those same advances in the development of consumer goods. We saw dramatic examples of technology spin-off as a result of the Apollo and Space Shuttle programmes. This can be expected to be repeated for the space station.

4. ARCHITECTURE

A space station concept includes a system of permanent, manned and unmanned elements in orbit, which communicate with ground support facilities. The concept also includes one or more unmanned, free-flying platforms, dedicated to commercial or scientific activities, orbiting near a manned base that would serve as a human habitat, utility core, laboratory, and orbital service station. The Shuttle Orbiter would carry the various station elements to orbit, help assemble the space station in orbit, and periodically return to the space station's manned element to bring supplies, fresh crews, and equipment. Transportation between the manned element and the free-flying platforms would be by a co-orbiting, unmanned space tug called an Orbital Manoeuvring Vehicle (OMV) based at the manned element. Eventually, the manned element would be the operations base for an Orbital Transfer Vehicle (OTV), a high energy, propulsive stage for transporting payloads to higher Earth orbits or into the solar system. This concept is shown in figure 1.

It is not yet certain what this space station system will look like because no space station design currently exists. However, N.A.S.A. is currently establishing the criteria for it. N.A.S.A would be very likely to adopt an evolutionary approach; that is, a space station capable of growing and changing to fit developing needs. N.A.S.A. would start with the design, development, test, and deployment of a small but useful space station capable of supporting a crew of six to eight in a low altitude, low inclination orbit. Figure 2 illustrates a few examples of possible configurations for a 1991 space station. Certain common architectural features can be noted. The laboratory modules, habitats, servicing bases, and other elements are all modules similar to the European Spacelab, the size of which is determined by the dimensions of the Orbiter payload bay. The Shuttle is the basic assembly and logistics vehicle.

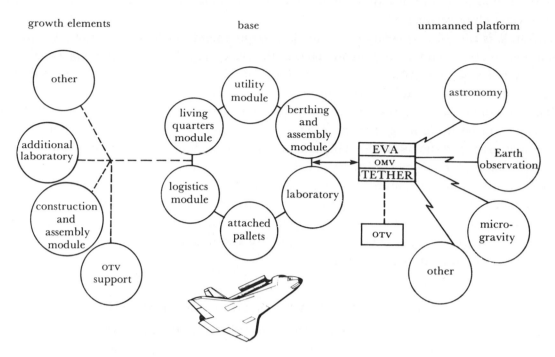

FIGURE 1. A space station architecture: cluster concept.

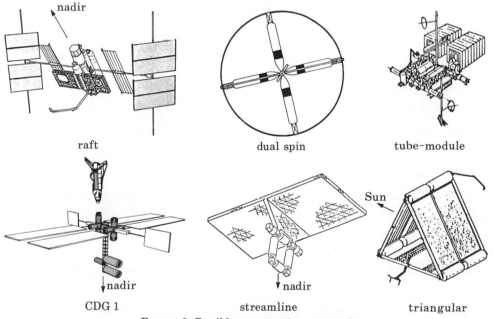

FIGURE 2. Possible space station geometries.

5. TECHNOLOGY AND OPERATIONAL NEEDS FOR A SPACE STATION

The space station development concept is based upon two radical approaches for spacecraft. One is that of evolving technology where systems can be replaced with advanced equipment as new technologies emerge. The second is the design for ease of maintenance and refurbishment.

These concepts should reduce developmental and operational costs, greatly enhance reliability, and obviate obsolescence.

N.A.S.A. is carefully assessing the technology requirements for a space station. Technology disciplines include data management, power, thermal, propulsion, and human capabilities. N.A.S.A. is developing a plan to move technology from its current level to that required by an evolving space station that will allow upgrading as technology matures.

Through these planning processes, numerous technological issues have been identified. The challenge will be to optimize the space station design so that the right level of autonomy in space operations and the proper level of maintainability is achieved, while still providing for evolving technology. The proper use and balance of automation and human capabilities will assure that the utility of the space station is brought to a level that can serve all user needs. The design, construction, and evolution of the space station must also assure low life cycle costs.

The use of dedicated systems such as Apollo can no longer serve our future purposes. Our planning must embrace the principles of reusability and maintainability yet provide technical flexibility to accommodate future requirements.

6. Status of U.S. planning for a space station

N.A.S.A's space station planning effort has been under way for almost two years. Many long-standing international partners have participated in these planning activities. This planning is consistent with President Reagan's policy 'to continue to explore requirements, operational concepts, and technology associated with permanent space facilities'. The planning guidelines are divided into management and engineering-related categories, as shown in table 2.

TABLE 2. Space station planning guidelines

management related	engineering related
three year extensive definition (5–10 % of programme cost)	continuously habitable dependent on the shuttle
planning throughout N.A.S.A.	
development funding in the fiscal year 1987	manned and unmanned elements
i.o.c. in 1991	evolutionary
cost of initial capability: $8 billion	maintainable
users involved in planning science and applications industry Department of Defense commercial	operationally autonomous user friendly
possible international participation	hardware can be upgraded as technology matures

The management-related guidelines include a comprehensive definition effort of approximately three years with an investment of 5–10 % of the total programme costs. The mission roles and responsibilities of the various N.A.S.A. centers will be determined early next year. N.A.S.A. is also conducting the necessary preparation and planning to support definition studies with industry in 1984. It is expected that several contracts will be let for different elements of the space station. Two or more contractors will conduct definition studies on the same element tasks. The programme is also being planned so that significant design and development funding would start in the fiscal year 1987 with a planned initial operational capability (i.o.c.) in orbit in the early 1990s. The cost of the initial capability is estimated to be about 8×10^9.

The engineering guidelines include provisions for continuous habitation, Shuttle support for initial launch, resupply and crew rotation, evolutionary growth through maintainable and restorable systems, manned and unmanned elements, and autonomous operation with the ability to upgrade systems as new technology becomes available. The entire system would be designed with emphasis on specific user needs.

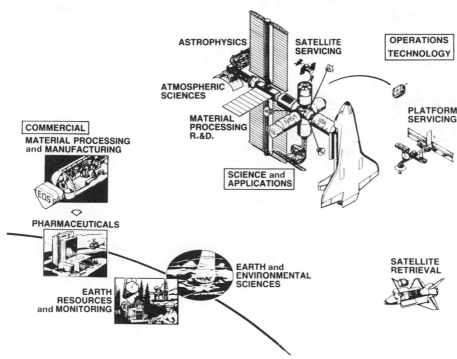

FIGURE 3. Initial concept for a space station.

An initial space station concept is shown in figure 3. It would satisfy initial user requirements by providing utilitarian laboratories for science and applications including commerical and defence applications complete with servicing, logistics, and operations support. It may also include special payload platforms and the capability to service them.

The initial capability is planned to evolve with emerging technology and advanced developments into the space station of the 2000s and beyond. A futuristic concept is shown in figure 4. It illustrates added capabilities to take payloads into higher orbits, service and repair satellites, retrieve various payloads, refuel vehicles, and many other capabilities to satisfy scientific, commercial, and security needs.

The planning schedule for the space station is shown in figure 5. All planning activities are geared toward initial operation in the early 1990s. The schedule includes planning activities to support the evolutionary process developing from the initial operational capability (i.o.c.) and leading up to the final operational capability (f.o.c.) in the year 2000.

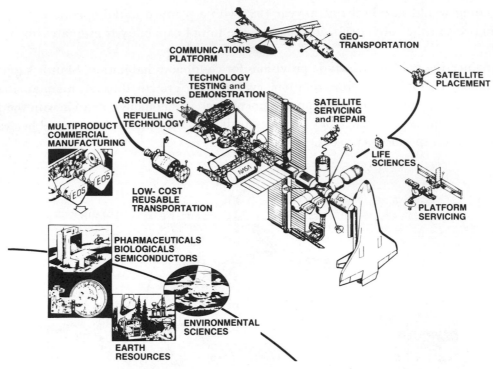

FIGURE 4. Future concept for a space station.

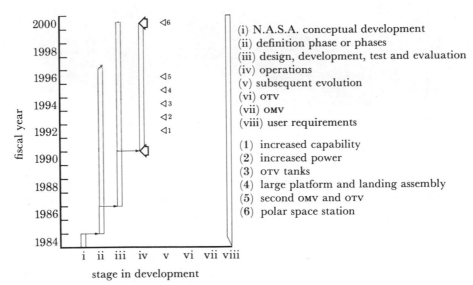

(i) N.A.S.A. conceptual development
(ii) definition phase or phases
(iii) design, development, test and evaluation
(iv) operations
(v) subsequent evolution
(vi) OTV
(vii) OMV
(viii) user requirements

(1) increased capability
(2) increased power
(3) OTV tanks
(4) large platform and landing assembly
(5) second OMV and OTV
(6) polar space station

FIGURE 5. Space station planning schedule.

7. THE USE OF A SPACE STATION

User requirements, which translate into functional capabilities, provide the foundation for the development of the space station architecture. The space station has focused on studies that emphasize user requirements for three groups: (i) science and applications; (ii) commerce; and (iii) technology development. Study contracts recently conducted by eight U.S. aerospace contractors assessed these requirements. The results of the study contracts were integrated into a single, time-phased mission set for a total of 107 missions. These are illustrated by discipline in table 3. On the basis of this mission set and the associated functional capabilities required of the space station system, the estimation of space station capabilities has been developed and is shown in table 4. These capabilities satisfy all initial requirements with the inclusion of two unmanned platforms, one co-orbiting with the manned space station and one in a near polar orbit. It should be noted that the future system requires 160 kW of power to the user as well as the space-based OTV for the transportation node function. As in the initial space station, materials processing facilities are the power drivers.

TABLE 3. PRELIMINARY DATA BASE FOR DEFINITION OF SPACE STATION MISSIONS (1991–2000)

science and applications	
astrophysics	22
Earth science and applications	5
Solar System exploration	10
life sciences	6
materials science	4
communications	1
subtotal	48
commercial	
materials processing in space	11
Earth and ocean observations	3
communications	14
subtotal	28
technology development	
materials and structures	5
energy conversion	3
computer science and electronics	4
propulsion	2
controls and human factors	5
space station systems and operations	8
fluid and thermal physics	4
subtotal	31
total	107

TABLE 4. ESTIMATION OF CAPABILITIES

	initial	future
base at 28.5°		
crew	6–8	12–18
power	75 kW	160 kW
attached payloads	some	more
servicing capability	initial (near)	better
orbital manoeuvring vehicle	available	available
data system	300 Mbit s^{-1}	300 Mbit s^{-1}
utility module	one	more
laboratory module	one	more
logistics module	two	more
living quarters module	one	more
multiple berthing adapter	one	under study
platforms		
base at 28.5°; 15 kW	one	several
polar; 15 kW	one	one
space-based OTV	no	yes
manned polar station	no	under study

8. Commercialization of space

The space station concept of today encompasses many new and broader spacecraft capabilities than were possible a decade ago. In addition to the science and applications laboratory functions, today's concept includes promising commercial opportunities. Commercial opportunities in space are defined as those products, processes, and services that demonstrate technical, economic, and institutional viability sufficient to achieve private sector investment in, ownership of, and operation of the activity as a profit-making venture.

The first 25 years of the space age has seen very significant programmes of commercial use of space systems for profitable enterprise. The dramatic achievements of the communications industry's exploration of geostationary satellites are well known. In the 1960s, N.A.S.A. undertook the development of the first communications satellites. Since that time, a vast, multinational business has emerged based on the idea of providing long distance communications services on Earth through exploitation of the global capabilities of communications satellites in synchronous equatorial Earth orbits. The tax revenues collected by the U.S. government each year from the corporations involved in this very profitable, space-based business exceed the total early N.A.S.A. investment in communications satellites. A similar success story can be told for the N.A.S.A. investment in meteorological satellites from which daily space-based photography has become a part of the nightly television weather report. The U.S. Navy's Transit navigation satellite is also a great commercial success with shipboard systems installed on over 4000 ships flying the flags of fifty nations. Even more significant is that these shipboard systems are manufactured by about fifteen nations. Similar steps are beginning for the commercialization of Earth observation systems such as the U.S. Landsat, the French Spot system, and the German MOMS equipment.

Likewise, the provision of launch services, initially government subsidized, is moving into commercialized operations with upper stages, launch boosters, and spacecraft being furnished by private industry.

N.A.S.A. has met with receptive firms in various industries to introduce the attributes of space and to start dialogues. These discussions have stimulated forward-looking industry managers, planners, and research leaders to conceive activities that can lead to profitable ventures in space. N.A.S.A. would then provide assistance and access to space technology with a view toward early proof of concepts, typically in N.A.S.A. low-gravity facilities such as drop tubes and towers, KC-135 flights and, in some cases, Shuttle flights. This programme has begun to stimulate innovative thinking in firms in a number of fields (for example, crystal growth, biological separations, containerless processing of alloys and other materials, and production of optical glass). The focus is on processes that gain a special advantage from the space environment and on products with high market value.

Two examples of commercial ventures are the McDonnell Douglas–Johnson and Johnson venture and the Microgravity Research Associates venture. McDonnell Douglas, using its own risk capital, is experimenting with a continuous-flow electrophoresis system in space, which is capable of yielding 700 times the quantity of certain kinds of pharmaceuticals with five times the purity of that available on Earth. Their market potential is enormous.

Microgravity Research Associates (M.R.A.) is a small, risk capital firm that will use its own funds to design and build an electroepitaxial furnace to produce gallium arsenide crystals that are virtually free of defects and are of very high purity for use in electronics and computer applications. M.R.A. believes that this kind of gallium arsenide crystal can be produced only in the microgravity of space. This material could potentially replace silicon as the microelectronic chip material of the future.

A space station, as N.A.S.A. now contemplates, will, on a long-term basis, accommodate many of the commercial processes and product applications now being studied by industry. It fosters further activity by enabling the development of space technology and operational techniques in fields such as space construction, Earth and space observation and monitoring, satellite servicing, testing, transportation node operation, and human endurance studies.

The commercialization of space in all its forms will have increasing implications for the U.S. economy, security, and world leadership and prestige. We have observed the value to Japan's economy gained through government and industry partnerships. Space offers one of several opportunities for government to demonstrate leadership and support to industry. There is a challenge to both sectors to use vision, creativity, and perseverance to make such a partnership work.

9. POTENTIAL FOR INTERNATIONAL PARTICIPATION

N.A.S.A. has seen a tremendous interest in its space station planning activities by many of its long-standing, international cooperative partners. While N.A.S.A. and the international space community recognize that the potential exists for international cooperation, all parties clearly understand that the space station programme has not been approved and that N.A.S.A. cannot and has not made any commitments at this time. Still, the growing space ambitions of the European Space Agency (E.S.A.), Canada, Japan, France, Germany, and Italy have prompted these countries to conduct parallel mission analysis studies at their own initiative and expense. Results of some of these studies compared with United States requirements are shown in table 5.

TABLE 5. SUMMARY OF INTERNATIONAL USE OF SPACE

mission area	U.S.	E.S.A.	Japan	Canada
material science and space processing				
manned R.&D. lab	×	×	×	×
attached processing facility	×			
co-orbiting processing facility	×	×	×	×
life science				
manned R.&D. lab	×	×	×	×
co-orbiting research facility				
space sciences and applications				
Earth observation				
high inclination free flyers and platforms	×	×		×
attached research facility	×			
manned high inclination platforms			×	
astronomical observation				
attached observatory	×			
low inclination free flyers and platforms	×	×	×	×
high inclination free flyer and platforms	×	×	×	×
technology and operations				
large structures	×	×	×	×
energetics	×	×	×	×
sensor development	×			×
maintenance, service and repair	×	×	×	×
OTV	×	×		×

Periodic meetings have been held to exchange study results. In addition, these countries are examining potential contributions to a space station if the opportunity to cooperate arises.

These early discussions are important for N.A.S.A. and its potential partners. Since the space station is viewed as a working orbital facility, N.A.S.A. is gaining information on potential world wide space station use patterns. Given the scope and complexity of a space station, this early involvement has given N.A.S.A. and the international community time to carefully evaluate cooperative possibilities should the decision be made to proceed.

N.A.S.A. has a long, successful history of cooperative activities. It has undertaken numerous activities with the E.S.A. and with individual European countries. For example, N.A.S.A. and British government agencies have participated in the Ariel satellite programme. Five satellites were launched from 1962 to 1974 to investigate a broad spectrum of scientific questions in the radio sciences, particles and fields, and high energy astronomy areas. Other cooperative activities include British hardware contributions to U.S. meteorological satellites, the N.A.S.A. International Ultraviolet Explorer, and the Solar Maximum Mission. More recently, the United Kingdom and the Netherlands participated in the Infrared Astronomical Satellite (I.R.A.S.), by providing the ground operations systems and an experimental instruments on the spacecraft. As we have all recently heard, I.R.A.S. has made some outstanding discoveries that astronomers will be studying for years. Finally, the United Kingdom is participating in the International Satellite-aided Search and Rescue Programme which has already demonstrated its ability to save lives.

In the field of space commercialization, several United Kingdom contractors have made successful, profitable arrangements with international industrial partners in the fields of communications satellites.

Additionally, Europe became a partner with N.A.S.A. in its Shuttle programme by building Spacelab. The recent launch of Spacelab aboard STS-9 marked the culmination of many years of hard work and excellent cooperation and is only the beginning of the possibilities that Spacelab provides. Spacelab offers numerous opportunities in several commercial areas and for science and applications, materials processing, and communications.

These joint activities have benefited all of us because we have been able to share the costs of these undertakings, thus allowing us to do more work. Very importantly, the scientific results from these activities have been shared among us.

It is our observation that large, peaceful, joint high-technology programmes among Western nations hold the hope of vitalizing high-technology industry on both sides of the ocean and strengthens our business and political ties. The only such programme in sight now and for the immediate future is the space station programme.

N.A.S.A's space station planning envisages a continuation of these policies of cooperation on space station development and use. There remains, however, the necessity of gaining a full understanding of cooperative and competitive activities and where one ends and the other begins. We enjoin all concerned to study the problem with us to ensure that we develop the right approach and a complete understanding on both sides.

10. Conclusion

The United States must continue space research, development, and applications to reap benefits for society and maintain space leadership. This continued involvement, shared with other nations and in cooperation with commercial companies, will expand the world's space activities into new realms. Just as we have shifted from pure space research and exploration in the 1960s toward applications in the 1970s and routine operations in the 1980s, space activities in the 1990s and into the next millennium will take on new meaning to mankind.

Just as N.A.S.A. had no idea that the Apollo programme would have so many benefits for society in the 1980s, N.A.S.A. cannot fully predict the impact and benefits of the space station in the year 2000. What is known is that the U.S.A. and many countries now have both a taste for space-borne knowledge and the institutional mechanisms for using new knowledge and rapidly converting it to new applications that come in many forms.

The 'high road' to space will soon be travelled by the common man, not just by those who possess 'the right stuff'. When the space station passes from the esoteric concepts of 1983 to the mundane activities of 2001, with factories in orbit, with permanent laboratories routinely yielding new knowledge, and with payload launches from the space station to the Moon, Mars, and other orbits in between and beyond, mankind will have found his place in the universe and will no longer be restricted to the world under his feet. We will then have the start of 'a city in space'.

Discussion

I. CRAWFORD (*University of Newcastle, U.K.*). What thought has been given to international financing of space stations?

R. F. FREITAG. N.A.S.A. has a long history of successful international cooperation in space activities. For space station studies conducted in 1982–1983, interest in potential cooperation has been expressed by E.S.A. and the Governments of Canada, France, West Germany, Italy and Japan. N.A.S.A. has welcomed their study participation but no commitments on cooperation have been made.

One groundrule that has been consistently followed in international cooperative projects is no exchange of funds. That is, each country funds its part of the mission. A good example in the space transportation system is the funding of the development of Spacelab by E.S.A. In the space station programme, we plan to follow the same policy.

Note added in proof (26 *March* 1984). President Reagan directed N.A.S.A. in his State of the Union message on 25 January 1984 to develop a permanently manned Space Station and to do it within a decade.

Phil. Trans. R. Soc. Lond. A **312**, 133–140 (1984)   [133]

Printed in Great Britain

The commercial potential of large orbital space platforms

By I. V. Franklin

British Aerospace Dynamics, Space and Communications Division (Bristol Site),
P.O. Box 5, Bristol BS12 7QW, U.K.

The commercial potential of large orbital space platforms or space stations is a key factor that must receive serious attention during the early feasibility and concept study phases. It is recognized that the total development cost will be very great and thus it is essential to identify potential users and their likely investment interest at an early stage. However, the development time of a space platform, say ten years, poses clear problems to both builders and potential users. Some typical problems are mission cost, operational effectiveness, commercial security, ownership and operational responsibility of the platform. Another very important issue is whether the platform should be manned, man-tended or unmanned. All these points are important factors to the commercial potential. The paper will review some of these aspects by using results gained during current studies being made for the European Space Agency.

Introduction

Space operations including spacecraft, launch vehicles and the ground segment are already in their third decade and a number of major milestones have been passed.

(i) The potential of the unmanned satellite and space probe is being fulfilled in a variety of roles such as space science, communications and Earth observation.

(ii) The highly successful commercial use of communications satellites is accepted.

(iii) Manned operations in space are now becoming almost routine after the success of the Space Shuttle.

(iv) The commercial potential of Earth observation satellites is now being developed.

(v) The ability to place and retrieve very large payloads into low Earth orbit (l.E.o.) has been demonstrated.

Most of these milestones have been achieved by using expendable launch vehicles such as Delta, Atlas Centaur, Saturn and Ariane. Perhaps the most significant achievement is the reusable launcher system known as the Space Shuttle, which has been demonstrated so successfully. This development has encouraged the consideration of space activities and related facilities for the next decades. Of particular interest is the study of the potential and architectural concepts of the manned space station and large orbital space platforms.

The space station concept envisages the following facilities: a permanent manned laboratory in space for scientific research in astronomy, Earth observation and life sciences; a base for manufacturing activities in space, which would exploit microgravity for the production of new materials for use on Earth; a transport node for spacecraft destined for high Earth orbit, which would integrate spacecraft and upper propulsion stages, and provide a base for an orbital transfer vehicle that would enhance the use of the Shuttle launch capacity. As a consequence, the Space Station is not seen simply as an updated version of the Skylab orbital laboratory but as a complex of components, some of which form the 'core' manned facility and some which form related elements. If the U.S. programme is approved it would be expected to develop

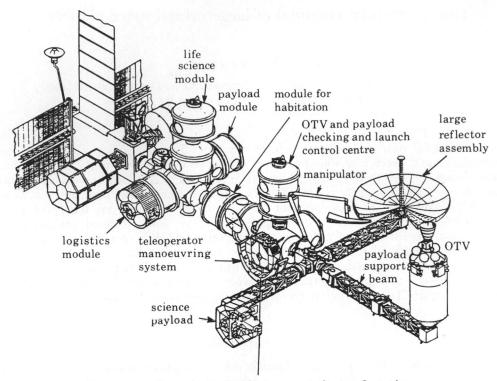

FIGURE 1. Concept of a baseline space station configuration.

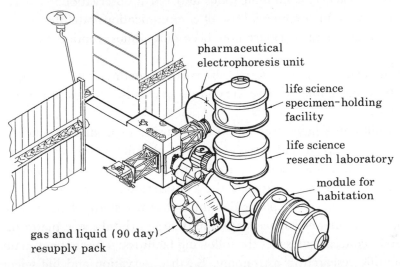

FIGURE 2. Initial space station with minimal logistics capability.

and evolve throughout the decade of the 1990s, beginning with a minimal four person facility in a 28.5° low Earth orbit in 1991. These major components will probably include: a baseline manned platform; a free-flying teleoperator servicing vehicle; a modular unmanned platform with space station elements but in other orbits (for example, polar); a reusable space based cryogenic orbital transfer vehicle (OTV); propellant storage and refuelling facilities and a vehicle servicing hangar associated with the manned station.

At this comparatively early stage it should be recognized that there is no definitive space station design but a number of concepts are being used for system assessment purposes. A typical concept that is being used in a European study (to be described later) is shown in figure 1. This concept has many of the features being used in the N.A.S.A. and U.S. industrial studies. Each of the basic elements or units shown in this concept can be carried into low Earth orbit by using the standard Space Shuttle, which gives a rough indication of the station size. The station will be assembled in space by using robotic and astronaut operated manipulators working out of the Space Shuttle. Figure 2 shows a concept of a much simpler platform, which can be unmanned or man-tended from Shuttle visits. These two basic concepts lead to the point that a space station can be manned, unmanned or man-tended, depending on the type and scale of mission to be undertaken.

A very significant milestone is the acceptance that space activities and operations should now offer a useful commercial potential whenever possible.

Space Station characteristics and environment

In all novel concepts and their working environment there are both advantages and disadvantages and some of the more important ones are given in table 1.

TABLE 1. Advantages and disadvantages of space station concepts

advantages	disadvantages
access to the unique environment of space for a wide range of users and equipment	high development costs and long development programme
a much greater payload capability in both mass and volume than can be provided by an unmanned automatic satellite	considerable crew safety constraints
flexibility of mission types and operational modes	potential users do not have control of the availability of the space transportation system
opportunities for non-astronaut participation in space activities for example, scientists, engineers and other technicians	the need to maintain independent commercial security for industrial users on a common space station
comparatively low recurring costs when compared to unmanned satellites	international political situations can have a serious influence on users' programmes
very long lifetime of basic hardware, together with the capability to regularly replenish consumables such as fuel, food and water	inevitably complex financial arrangements for development and use, particularly when considering international involvement
the ability to repair and replace worn or damaged major components	
the ability to make 'hands on' experiments and processing using man's unique abilities for subjective judgement, tactile skills and response to unusual situations	

Why put people in space?

In this age of highly developed automation and robotic operations in a wide variety of industrial activities, it is proper to ask the question, 'why put people in space?'. The point is that man should only perform tasks that use his unique skills and not be used in functions that machines and automation can do better. Some examples of such abilities are: the ability to make real-time subjective decisions based on direct observation and the facility to deal with the unexpected; that man can provide the application of continuous intelligence with an inbuilt adaptive logic and programming ability; the ability to assess situations and communicate.

The space environment

It is appropriate to include some of the more significant characteristics of the space environment that have relevance to the operation of a space station. It is not often realized that space is a comparatively benign and predictable environment. There is no climate, no bad weather, no natural disasters of the type experienced on Earth.

Some of the characteristics are: a continuous microgravity level of 10^{-3} to $10^{-5}\,g$; a continuous hard vacuum; a solar and ultraviolet radiation not subject to atmospheric attenuation; a unique experiment base for work that is impossible in a normal terrestrial environment; the capability to provide facilities for the remote observation of Earth at all latitudes and longitudes, i.e. global coverage; free solar energy, which after photovoltaic conversion by solar arrays can provide the station's electrical power supply.

Related disciplines and potential users

Most potential users may be expected to be identified within the following broad disciplines: materials science and associated processes; life sciences; space sciences; environmental observation from space; space technology and testing; satellite repair and replacement and mission support services.

TABLE 2. The disciplines in materials science and life science that could use the space environment

metal composites
    solidification of metal melts, also with dispersed additives
    metal foams
    measurement of physical parameters
    improvement and testing of technologies in space and on Earth

interface and transport phenomena
    basic study of cellular and Marangoni convection
    studies of boundary and transport phenomena at and in interfaces and surfaces
    measurement of transport parameters

crystals
    basic study of melt zone crystal growth
    growth of new types of semiconductors
    growth of monocrystals with finely dispersed inclusions
    growth of crystals by precipitation from solution
    developments in glass, ceramics and refractories

physical chemistry and process engineering
    measurement of thermal and caloric state functions
    studies on reaction kinetics
    containerless processing
    measurement of physical parameters
pharmaceuticals and bioprocessing

life science

human physiology and medicine	radiation biology
gravitational biology	exobiology
	biotechnology

Preliminary surveys have indicated that initial potential users of space station or platform facilities are in the many subdisciplines of material and life sciences. These interests are a direct consequence of the inherent availability of the microgravity environment. An indication of the range of these interest for materials science is given in table 2. The lists in table 2 show that the potential commercial users can be expected to be found in the companies responsible for producing very high quality metals, crystals, ceramics and pharmaceuticals. It is stressed that there is no intention of proposing materials processing in space for products that can be produced simply on Earth; it is reserved for the exploitation of processes and materials that will benefit significantly from manufacture in space. Indeed, the current situation in the U.S. is that a major pharmaceuticals house, Johnson and Johnson, in close collaboration with McDonnell Douglas, have already flown a continuous flow electrophoresis experiment on four Shuttle flights with very encouraging results. The results have shown that an increase in yield of over five hundred times has been achieved with quantitatively repeatable separation. The flights have also demonstrated the value of manned participation as well as validating the design concepts.

The main driver for the interest in materials science is the very low gravity environment known as microgravity, although the actual levels lie between $10^{-3}\,g$ and $10^{-5}\,g$ depending on the operational requirements and facility. Microgravity leads to some interesting physical and engineering phenomena such as convection free conditions in liquids and gases, and no gravity-dependent separation of materials with large differences in density. These two characteristics can be used for the production of more perfect homogeneous materials or materials that cannot be produced under normal Earth gravitational conditions. Continuous flow electrophoresis is a typical process that benefits dramatically by the elimination of gravity.

EUROPEAN POTENTIAL USERS

As part of its Long-Term Preparatory Programme (L.T.P.P.) on space transportation systems the European Space Agency (E.S.A.) has identified a number of areas for study by selected European industrial and establishment teams. Of particular interest is the recently completed first phase of one of those studied, which is called 'European Utilisation Aspects of a U.S. Manned Space Station', a title that has been abbreviated to EUA-1. This E.S.A. study was led by D.F.V.L.R., the German Space and Aeronautics Research Establishment with an industrial team comprising ERNO, Aeritalia, Matra, Dornier System and British Aerospace, all of whom have been involved in Spacelab. The main objectives of the study were to identify the level and type of European interest in using a manned space station; to identify potential European payload candidates, which will be beneficially supported by a manned space station, and to assess the required operational support from the Station; to discuss alternative approaches and identify the impact if no manned space stations were available to support the missions.

The plan was for each industrial company to make personal contact with companies and institutions working in the relevant fields to establish the level of interest in the potential offered by the space station. The contact was followed up with a requirements questionnaire and a short descriptive brochure of a typical space station and its facilities. From these contacts and questionnaires it was hoped that a reasonably well defined set of candidate payloads and their requirements from the station would emerge.

This approach identified a fundamental problem. It was a comparatively simple matter to work with scientists who were already familiar with space activities, but it was much more difficult to contact and communicate with potential users outside the space community. The problem was bilateral; the understanding of what space could offer and what industry needed demonstrated a requirement for mutual education that could not be achieved quickly. This problem was particularly acute for the pharmaceuticals industry. It is worth noting that fundamental research into exotic metallurgy and crystallography in a microgravity environment has been proceeding for several years by using the very successful Skylark sounding rocket, which gives several minutes of microgravity, which is often sufficient for proof of concept tests. The programme, known as TEXUS, is strongly supported by the German D.F.V.L.R.

The first phase of the EUA study provided valuable lessons, particularly on user contact, which have been incorporated into the second phase of that study, which has recently started. It is also hoped that symposia of this type will help to bridge the gap between potential users and those who understand the space environment and can provide the facilities. However, despite the difficulties, some 35 materials science payloads were identified and, as had been expected, these were mainly in basic research. Twenty five of the proposed payloads were German. During the study it was observed that there was an interest by potential users in doing materials and processes research in orbit, but an understandable reluctance to speculate about the potential for large scale space production until the essential research is completed. Another viewpoint was that useful experiments could be done in space, which would lead to a better understanding of terrestrial processes. Many other payloads were identified in the other fundamental disciplines described earlier, but as they have less commercial potential at this time they will not be discussed further.

So, the basic objectives of the study were met, useful lessons were learned, which have been recognized in the second phase activities, and the broad summary and conclusions of the European study are now given.

Summary

(i) A manned space station is necessary for selected European payloads in material and life sciences.

(ii) Manned space station operations need to be complemented by free-flying unmanned platforms for automatic materials processing.

(iii) Most payloads identified were for basic research.

(iv) Commercial opportunities at this stage were only identified in communications, where the space station is used for orbit staging and assembly.

(v) Identification of possible materials-processing payloads will be largely dependent on the flight results obtained from the Spacelab mission.

Conclusions

Participation in a U.S. manned space station programme will promote European scientific and technological progress. The usefulness of people for representative candidates from all user disciplines was identified. The identified trends in use result in increasing payload requirements to be met by future space systems in both technical and functional capabilities. Human assistance is important for the development of these capabilities even if in the long term they may be provided by automatic systems.

Participation in a U.S. manned space station will, in the long term, open the way for increased operational and commercial use of space by Europe. As well as a working research base, the space station can be a service centre, transfer station and warehouse supporting future space missions. It will simplify access to space and increase the cost-effectiveness of space operations, both of which are of considerable importance to the successful commercial use of space.

COMMERCIAL POTENTIAL

The industrialization of space, and its related commercialization, has already started; it is no longer science fiction but is rapidly becoming science fact. As other papers have shown, the communications industry, both suppliers and users of equipment, were quick to realise the great commercial benefits that fast, high quality, high capacity space communications could offer by using satellites. In this business sector, the use of space facilities is no longer a novelty. Although space station studies are at a comparatively early stage, there is a growing feeling that there is great scope for their commercial potential. In this context commercial potential should be associated with cost benefits. The absolute assessment of cost benefits is extremely difficult and will not be attempted here, but it is possible to indicate trends and growth areas. In this third decade of space activities it is clear that unless the space station can demonstrate a significant commercial potential, then its future, at least for civil applications, may be limited.

It is important to understand that cost benefits or commercial returns are not immediate and will require those who wish to invest in such a large scale advanced programme to accept that the development of an operational manned space station will take about nine years. There is no quick return on investment, which is true of most high technology industry. To develop and maintain the interests of potential users it is essential to demonstrate the ability of space facilities to provide the services required. Commercial growth and product development will be strongly dependent on the results to be obtained from early flights, which will be concentrating on research and development. In Europe, for example, much depends on the success of Spacelab, which is being flown in the Space Shuttle. This is a typical example of the type of space laboratory that is an essential tool for space processing experiments in both science and technology. The current experience of McDonnell Douglas and Johnson and Johnson in their continuous-flow electrophoresis equipment for achieving very high purity production indicates a time of five to ten years to develop from concept to implementation, for the introduction of a new product to the commercial market.

It is recognized that the use of space systems in an industrial process will inevitably carry a risk, particularly for commercial ventures, and the space industry must provide a continuing public relations service to encourage potential users. Further, very important points are concerned with protection of proprietary rights, intellectual property rights, access to the facilities, the potential for private ownership, and crew training for company operatives working on the station. To encourage commercial interest for space-developed products and services, space facilities must be easily accessible by dependable and regularly scheduled space transportation systems with a clearly defined tariff.

Economic benefits of research and development are relatively difficult to quantify. One method of evaluating these benefits is to consider a function such as the 'cost per kilogram hour', which is based on two factors, namely payload capability and mission duration. Some figures have been estimated in the U.S. space station studies. With the basic assumption that the space station can operate on a 90 day production cycle on materials processing, the costs have been estimated at about $2.00 per kilogram hour, as compared with $17 per kilogram hour for Space Shuttle processing operations and very much higher costs for other methods of materials processing in space. However, at present this is a highly speculative area and much will depend on the outcome of the relevant Shuttle flights.

One U.S. estimate of space station economic benefits, made in 1983, is $1.6 billion and does not include any income or benefits from commercial materials processing in space. The benefits come from the use of a space-based orbit transfer vehicle (69 %), satellite servicing (14 %) and research and production (17 %). For all three categories the space station mission costs were compared with both the economic value of the mission (when quantifiable) and the projected cost of alternative means for accomplishing the same mission. Conservative estimates were used throughout. The potential benefits of materials processing in space are believed by some experts to be of the order of billions of dollars per year but much more evidence is required before these claims can be substantiated.

The mass of European microgravity payloads transported into low Earth orbit in the decade beginning 1995 is likely to increase owing to the availability of orbital facilities such as Eureca, the European retrievable carrier, and the manned space station, and a growing awareness of the commercial potential for space materials processing. At this time an analysis of experimental payloads identifies something like 6.5 t of materials science payloads and 5.6 t of life science payloads. How many will actually fly depends on funding availability, but equally as facilities become available and experiments are successful the demand might be expected to grow.

Two factors may limit the rate of growth of microgravity payloads in the post-1995 period. The first is the time delay between the inception and completion of an experiment indicated earlier, and the second is the specific cost of material processed in a space facility, which may limit its commercial exploitation. Several iterations may be required to bring a commercial process to maturity. The specific cost of processed materials determines which products might show commercial attractiveness. The relation between specific cost and demand may be as powerful as an inverse cube relation, so that cost reduction in this period may be very important indeed.

CONCLUSIONS

At this early study stage it is not possible to give many firm conclusions, but certain points have emerged from both the U.S. and European studies.

(i) Candidate commercial users do exist, but it is essential that there is mutual involvement and continuing dialogue to sustain and foster the interest.

(ii) There is a gap between potential users and the space industry and strong efforts must be made to bridge this gap.

(iii) A manned facility is essential for research and development, operations and commercialization.

(iv) The manned space station must be evolutionary and affordable, and benefits should repay investment.

(v) Management aspects, both commercial and operational, need careful consideration.

(vi) The ultimate space station system must include both manned and unmanned elements in various orbit locations.

The author wishes to thank British Aerospace Dynamics, Space and Communications Division, for permission to publish this paper, and in particular the support given by Dr R. Parkinson. Thanks are due to the Headquarters of the European Space Agency for permission to use the results of the recently completed study. Finally, thanks are due to McDonnell Douglas Astronautics for their considerable help in providing some of the illustrations used.